Machine Design Data Book

Second Edition

Machine Design Data Book

Second Edition

Compiled by

Dr. P. V. Ramana Murti

Retired Professor and Director of Evaluation
JNTU, Hyderabad.

Dr. M. Vidyasagar

Professor of Mechanical Engineering
JNTUH College of Engineering, Hyderabad.

BSP BS Publications

A unit of **BSP Books Pvt. Ltd.**
4-4-309/316, Giriraj Lane,
Sultan Bazar, Hyderabad - 500 095

Machine Design Data Book *by*

P. V. Ramana Murti and M. Vidyasagar

Published by :

BS Publications
A unit of **BSP Books Pvt. Ltd.**
4-4-309/316, Giriraj Lane, Sultan Bazar,
Hyderabad – 500 095.
Phone : 040 – 23445600, 23445688
e-mail : info@bspbooks.net
www.bspbooks.net

ISBN: 978-93-5230-111-9

Preface to Second Edition

In this revised edition, a topic 'Fibre Ropes' is included in the chapter on 'Power Transmission Systems, Pulleys'. Also an additional chapter on 'Shafts and Axles' is added and this covers the JNTUH Syllabus for 'Advanced Machine Design' subject in M.Tech Engineering Design course.

-Authors

Preface to First Edition

Design of machine members is a basic subject in Mechanical Engineering. It is offered as two courses in many universities. The first course deals with general principles of design viz. design of shafts, joints and couplings for which design data handbooks are not suggested. The second course deals with design of machine members like belts, pulleys, bearings, gears, levers and I.C. Engine parts etc., which require the use of handbook and the universities permit handbook during the examination. The authors who had lot of experience in teaching the subject felt that though there are excellent voluminous data handbooks available, just meeting the needs of the topics in second course on design of machine members might be helpful. Keeping this in mind, the data has been compiled and data book prepared for the purpose of class room teaching and examinations in the subject of second course on design of machine members. S.I system of units has been followed throughout the handbook. While compiling data, the support of information presented in several textbooks, data books and IS standards is made use of and the same is acknowledged. The authors are thankful to the authorities of JNTUH University, Hyderabad for their cooperation and giving permission to publish the data book. They are also thankful to the faculty of Mechanical Engineering Department in JNTUH College of Engineering and specially thankful to Dr. A. Seshu Kumar, Senior Principal Scientist, IICT, Hyderabad and Prof. M. Sreenivasa Rao, Department of Mechanical Engineering, JNTUH for their constant support and continuous encouragement at various stages during the preparation of this data book. The authors are thankful to BS Publications for their excellent cooperation. Suggestions and comments for improvement of the data book will be appreciated.

- *Authors*

Contents

Chapter 1

Sliding Contact Bearings

McKee's equation, $\mu = \left[\frac{33.25}{10^8} \times \frac{Zn}{p} \times \frac{d}{c}\right] + k$

μ – Coefficient of friction

Z – Absolute or dynamic viscosity of oil, N-s/m^2 or kg/m-s

n – Speed of journal, rpm

p – Bearing pressure, N/mm^2

d – Diameter of journal, mm

c – Diametral clearance, mm

k – 0.002

Sommerfeld number, $S = \frac{Zn}{60 \times 10^6 p}\left(\frac{d}{c}\right)^2$ (Refer Table 1.1)

$$p = \frac{W}{l d}$$

W – Load acting on the journal, N

l – Length of journal, mm

Heat generated in the bearing, $H_g = \mu W V$ watts

V – Sliding velocity of journal, m/s

Heat dissipated by the bearing, $H_d = CA\ (t_b - t_a)$ watts where

C – Heat dissipation coefficient

= 140 to 420 W/m^2/°C (still air)

= 490 to 1400 W/m^2/°C (well ventilated bearings)

A – Projected area ($l \times$ d) in m^2

t_b – Temperature of bearing surface, °C

$$t_b = \frac{t_o + t_a}{2}$$

t_o – Temperature of oil, °C

t_a – Atmospheric temperature (temperature of surrounding air), °C

Mass of oil required to carry away the heat generated, $m = \frac{H_g}{C_p \Delta t_o}$ kg/s

where

C_p – Specific heat of oil

= 1840 to 2100 J/kg/°C

Δt_o – Difference between outlet and inlet temperatures of oil, °C

Critical pressure (i.e., minimum operating pressure)

$$p_c = \frac{Zn}{4.75 \times 10^6}\left(\frac{d}{c}\right)^2\left(\frac{l}{l+d}\right) \text{ N/mm}^2$$

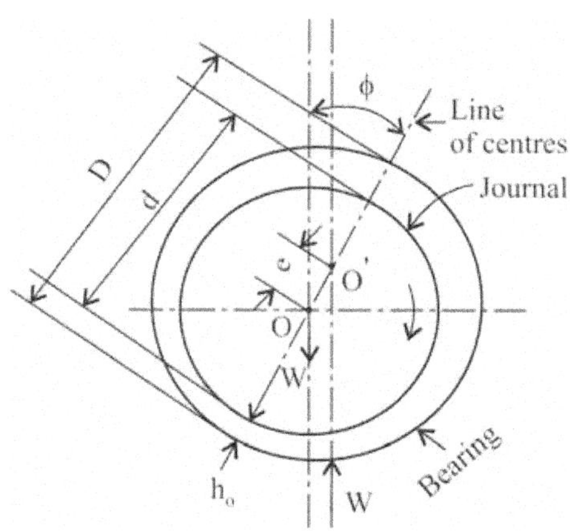

Fig. 1.1 Geometric relations for a journal bearing

Diametral clearance, $c = D - d$

Diametral clearance ratio, $c_r = \frac{c}{d}$ ($\simeq 0.001$ usually)

Eccentricity, $e = \frac{c}{2} - h_o$

Minimum film thickness, $h_o = \frac{c}{2} - e = \frac{c}{2}(1 - \in)$

Eccentricity factor or attitude, $\in = \frac{2e}{c} = 1 - \frac{2h_o}{c}$

Dimensionless performance parameters:

$\frac{\mu d}{c}$ – Coefficient of friction variable

$\frac{4q}{dcn'l}$ – Flow variable

$\frac{q_s}{q}$ – Flow ratio

$\frac{p}{p_{max}}$ – Pressure ratio

$\frac{\rho C_p \Delta t_o}{p}$ – Temperature rise variable

where

q – Oil flow through the bearing, mm^3/s

q_s – Axial flow of oil (or) side leakage, mm^3/s

n' – Speed of journal in rps

p_{max} – Maximum bearing pressure, N/mm^2

ρ – Density of oil = 900 kg/m^3

$\rho C_p = 142 \times 10^4$ N/m^2 °C

ϕ – Attitude angle

Table 1.1 Dimensionless performance parameters for full journal bearings with side flow

$\frac{l}{d}$	ϵ	$\frac{2h_o}{c}$	S	ϕ	$\mu\frac{d}{c}$	$\frac{4q}{dcn'l}$	$\frac{q_s}{q}$	$\frac{\rho C_p \Delta t_o}{p}$	$\frac{p}{p_{max}}$
	0	1.0	∞	70.92	∞	π	0	∞	-
	0.1	0.9	0.240	69.10	4.80	3.03	0	19.9	0.826
	0.2	0.8	0.123	67.26	2.57	2.83	0	11.4	0.814
	0.4	0.6	0.0626	61.94	1.52	2.26	0	8.47	0.764
∞	0.6	0.4	0.0389	54.31	1.20	1.56	0	9.73	0.667
	0.8	0.2	0.021	42.22	0.961	0.760	0	15.9	0.495
	0.9	0.1	0.0115	31.62	0.756	0.411	0	23.1	0.358
	0.97	0.03	-	-	-	-	0	-	-
	1.0	0	0	0	0	0	0	∞	0
	0	1.0	∞	85	∞	π	0	∞	-
	0.1	0.9	1.33	79.5	26.4	3.37	0.150	106	0.540
	0.2	0.8	0.631	74.02	12.8	3.59	0.280	52.1	0.529
	0.4	0.6	0.264	63.10	5.79	3.99	0.497	24.3	0.484
1	0.6	0.4	0.121	50.58	3.22	4.33	0.680	14.2	0.415
	0.8	0.2	0.0446	36.24	1.70	4.62	0.842	8.00	0.313
	0.9	0.1	0.0188	26.45	1.05	4.74	0.919	5.16	0.247
	0.97	0.03	0.00474	15.47	0.514	4.82	0.973	2.61	0.152
	1.0	0	0	0	0	-	1.0	0	0
	0	1.0	∞	88.5	∞	π	0	∞	-
	0.1	0.9	4.31	81.62	85.6	3.43	0.173	343.0	0.523
	0.2	0.8	2.03	74.94	40.9	3.72	0.318	164.0	0.506
	0.4	0.6	0.779	61.45	17.0	4.29	0.552	68.6	0.441
0.5	0.6	0.4	0.319	48.14	8.10	4.85	0.730	33.0	0.365
	0.8	0.2	0.0923	33.31	3.26	5.41	0.874	13.4	0.267
	0.9	0.1	0.0313	23.66	1.60	5.69	0.939	6.66	0.206
	0.97	0.03	0.00609	13.75	0.610	5.88	0.980	2.56	0.126
	1.0	0	0	0	0	-	1.0	0	0
	0.0	1.0	∞	89.5	∞	π	0	∞	-
	0.1	0.9	16.2	82.31	322.0	3.45	0.180	1287.0	0.515
	0.2	0.8	7.57	75.18	153.0	3.76	0.330	611.0	0.489
	0.4	0.6	2.83	60.86	61.1	4.37	0.567	245.0	0.415
0.25	0.6	0.4	1.07	46.72	26.7	4.99	0.746	107.0	0.334
	0.8	0.2	0.261	31.04	8.80	5.60	0.884	35.4	0.240
	0.9	0.1	0.0736	21.85	3.50	5.91	0.945	14.1	0.180
	0.97	0.03	0.0101	12.22	0.922	6.12	0.984	3.73	0.108
	1.0	0	0	0	0	-	1.0	0	0

Table 1.2 Design practices for journal bearings

Machinery	Bearing	$\frac{l}{d}$	Allowable bearing pressure, p N/mm²	Absolute viscosity of lubricant, Z N-s/m²	$\left(\frac{Zn}{p}\right)_{min}$
Stationary High Speed Steam Engines	Main Crank Pin Wrist Pin	1.5 –3.0 0.9–1.5 1.3–1.7	1.75 4.2 4.6	0.015 0.03 0.025	3.56 0.85 0.71
Gas and Oil Engines (Four Stroke)	Main Crank Pin Wrist Pin	0.6 –2.0 0.6–1.5 1.5–2.0	4.9–8.4 10.8–12.6 12.5–15.4	0.02–0.065	2.85 1.42 0.71
Gas and Oil Engines (Two Stroke)	Main Crank Pin Wrist Pin	0.6 –2.0 0.6–1.5 1.5–2.2	3.5–12.5 7–10.5 8.4–12.5	0.02–0.065	3.56 1.7 1.42
Aircraft and Automobile Engines	Main Crank Pin Wrist Pin	0.8–1.8 0.7–1.4 1.5–2.2	5.6–11.9 10.5–24.5 16.1–35	0.008	2.13 1.42 1.14
Reciprocating Compressors and Pumps	Main Crank Pin Wrist Pin	1.0–2.2 0.9–1.7 1.5–2.0	1.75 4.2 7.0	0.03 –0.08	4.27 2.85 1.42
Centrifugal Pumps, Motors and Generators	Rotor	1.0–2.0	0.7-1.4	0.025	28.45
Machine Tools	Main	1.0–4.0	2.1	0.04	0.14
Steam Turbines	Main	1.0–2.0	0.7–2.0	0.002–0.016	14.22
Railway Cars	Axle	1.9	3.5	0.1	7.11
Marine Steam Engines	Main Crank Pin Wrist Pin	0.7–1.5 0.7–1.2 1.2–1.7	3.5 4.2 10.5	0.03 0.04 0.03	2.85 2.13 1.42
Transmissions	Light, Fixed	2–3	0.18	0.025	14.22
Gyroscopes	Rotor	–	6.0	0.03	7.82
Shafting	Self Aligning Heavy	2.5–4 2–3	1.1 1.1	0.06 0.06	4.27 4.27
Cotton Mills	Spindle	–	0.007	0.002	14.24
Punching and Shearing Machines	Main Crank Pin	1–2 1–2	28.0 56.0	0.1 0.1	– –
Rolling Mills	Main	1–1.5	21.0	0.05	1.42

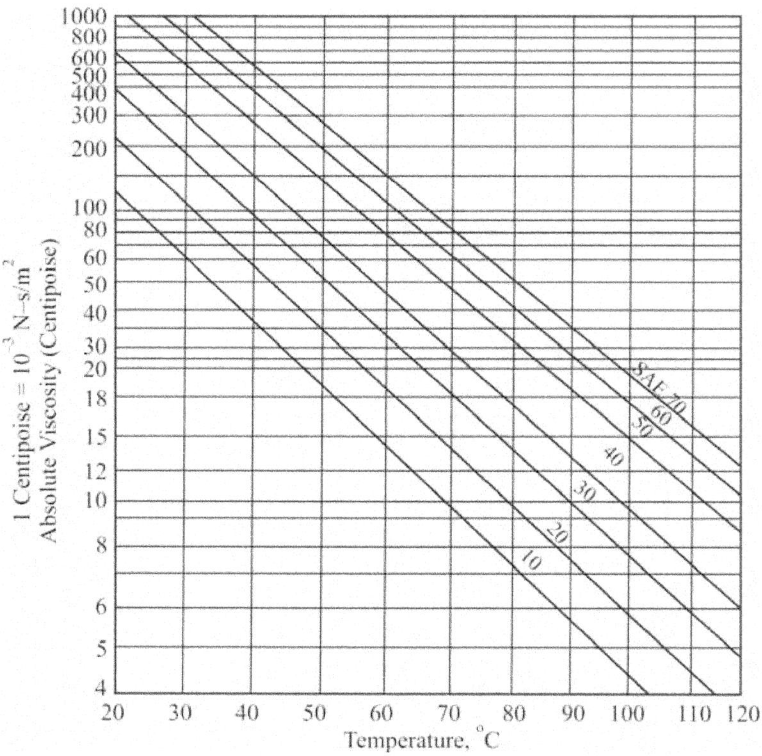

Fig. 1.2 Average absolute viscosities versus temperature

Chapter 2

Rolling Contact Bearings

Dynamic capacity, $C = \left(\dfrac{L}{L_{10}} \right)^{\frac{1}{k}} P$

where $k = 3$ for ball bearing

$= \dfrac{10}{3}$ for roller bearings

$L_{10} = 1$ million revolutions

Equivalent load, $P = (V X F_r + Y F_a) S$

where $V = 1.0$ for inner ring rotating and outer ring stationary

$= 1.2$ for inner ring stationary and outer ring rotating

F_r, F_a – Radial and axial loads

X, Y – Radial and axial load factors

(Refer Table 2.1)

S – Service factor

= 1.0 for constant or steady load

= 1.5 for light shocks

= 2.0 for medium shocks

= 2.5 for heavy shocks

Life of bearings in million revolutions (mr), $L = \left(\dfrac{C}{P} \right)^{k}$ mr

Life of bearings in hours, $L_h = \dfrac{L \times 10^6}{60 \times n}$ hours

n – Speed in rpm

Cubic mean load (P_m) for periodically variable loading

$$P_m = \left[\frac{P_1^3 n_1 + P_2^3 n_2 + \ldots P_x^3 n_x}{n_1 + n_2 + \ldots n_x}\right]^{1/3} \quad \text{(For revolutions varying)}$$

$$= \left[\frac{P_1^3 t_1 + P_2^3 t_2 + \ldots P_x^3 t_x}{t_1 + t_2 + \ldots t_x}\right]^{1/3} \quad \text{(For time varying)}$$

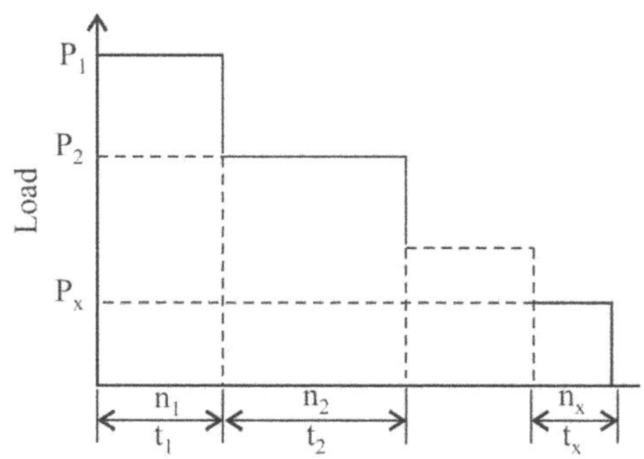

$n_1, n_2 \ldots n_x$ – Number of revolutions for the loads $P_1, P_2 \ldots P_x$

$t_1, t_2 \ldots t_x$ – Periods of rotation for the loads $P_1, P_2 \ldots P_x$

Probability of survival, $\dfrac{L}{L'_{10}} = \left[\dfrac{\ln(1/p)}{\ln(1/p_{10})}\right]^{1/b}$

where L – Required life of bearings in million revolutions

L'_{10} – Calculated life of selected bearings, for the given load, for 90% survival

p – Probability of survival

p_{10} – Probability of survival for 90% (or) 0.9

$$\ln(1/p_{10}) = \ln(1/0.9) = 0.1053$$

b = 1.17 for a median life L_{50} ($L_{50} = 5\,L_{10}$)

Table 2.1 Radial and axial load factors

Type of bearing	Series (SKF)		$\frac{F_a}{F_r} \le e$		$\frac{F_a}{F_r} > e$		e
			X	Y	X	Y	
Deep groove ball bearing C_0 = static capacity of bearing	Series 60, 62, 63, 64	$\frac{F_a}{C_0} = 0.025$	1	0	0.56	2	0.22
		= 0.04	1	0	0.56	1.8	0.24
		= 0.07	1	0	0.56	1.6	0.27
		= 0.13	1	0	0.56	1.4	0.31
		= 0.25	1	0	0.56	1.2	0.37
		= 0.5	1	0	0.56	1.0	0.44
Angular contact ball bearing	72B, 73B, 33		1	0	0.35	0.57	1.14
			1	0.73	0.62	1.17	0.86
Self aligning ball bearing	2200-2204		1	1.3	0.65	2	0.5
	05-07		1	1.7	0.65	2.6	0.37
	08-09		1	2	0.65	3.1	0.31
	10-13		1	2.3	0.65	3.5	0.28
	14-20		1	2.4	0.65	3.8	0.26
	21-22		1	2.3	0.65	3.5	0.28
	2301		1	1	0.65	1.6	0.63
	2302-2304		1	1.2	0.65	1.9	0.52
	05-10		1	1.5	0.65	2.3	0.43
	11-18		1	1.6	0.65	2.5	0.39
Spherical roller bearing	22205C-22207C		1	2.1	0.67	3.1	0.32
	08C-09C		1	2.5	0.67	3.7	0.27
	10C-20C		1	2.9	0.67	4.4	0.23
	22C-44C		1	2.6	0.67	3.9	0.26
Taper roller bearing	32206-208		1	0	0.4	1.6	0.37
	09-22		1	0	0.4	1.45	0.41
	24-30		1	0	0.4	1.35	0.44

Table 2.2 (a) Deep groove ball bearings

Bearing No. (SKF)	d (mm)	D (mm)	B (mm)	Basic capacity, N	
				Static C_0	Dynamic C
6000	10	26	8	1865	3550
01	12	28	8	2160	4000
02	15	32	9	2500	4400
6003	17	35	10	2800	4650
04	20	42	12	4415	7350
05	25	47	12	5100	7800
6006	30	55	13	7000	10400
07	35	62	14	8840	12500
08	40	68	15	9615	13200
6009	45	75	16	12460	16300
10	50	80	16	13440	17000
11	55	90	18	17660	22000
60012	60	95	18	18950	22800
13	65	100	18	21200	24000
14	70	110	20	25500	30000
6015	75	115	20	28000	31000
16	80	125	22	33500	37500
17	85	130	22	36000	39000
6018	90	140	24	41500	45500
19	95	145	24	45000	47500
20	100	150	24	45000	47500
6021	105	160	26	54000	57000
22	110	170	28	61000	64000
24	120	180	28	65500	67000
6026	130	200	33	83000	83000
28	140	210	33	90000	86500
30	150	225	35	104000	98000
6032	160	240	38	118000	112000
34	170	260	42	143000	132000
36	180	280	46	166000	150000
6038	190	290	46	180000	153000
40	200	310	51	200000	170000

Table 2.2 (b) Deep groove ball bearings

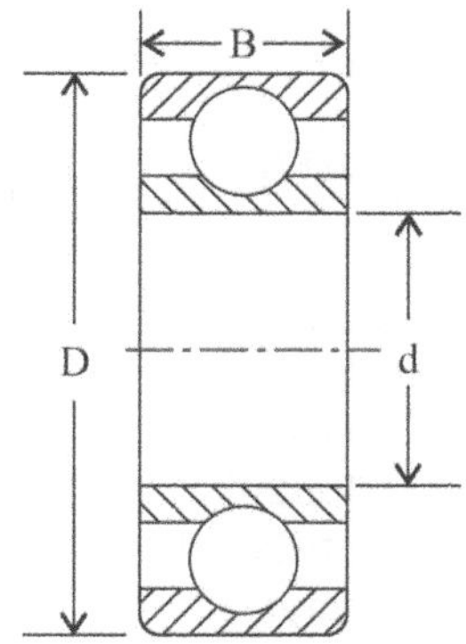

ISI No.	Bearing No. (SKF)	d (mm)	D (mm)	B (mm)	Basic capacity, N	
					Static C_0	Dynamic C
10BC02	6200	10	30	9	2160	3925
12BC02	01	12	32	10	2940	5250
15BC02	02	15	35	11	3430	5980
17BC02	6203	17	40	12	4315	7355
20BC02	04	20	47	14	6375	9805
25BC02	05	25	52	15	6965	10690
30BC02	6206	30	62	16	9805	14710
35BC02	07	35	72	17	13535	19615
40BC02	08	40	80	18	15495	22165
45BC02	6209	45	85	19	17750	24910
50BC02	10	50	90	20	20595	27070
55BC02	11	55	100	21	25300	33340
60BC02	6212	60	110	22	31580	40210
65BC02	13	65	120	23	35715	42905
70BC02	14	70	125	24	38440	47070
75BC02	6215	75	130	25	41385	50600
80BC02	16	80	140	26	44130	55505
85BC02	17	85	150	28	53450	63745
90BC02	6218	90	160	30	60800	74040
95BC02	19	95	170	32	71100	82870
100BC02	20	100	180	34	79925	97145
105BC02	6221	105	190	36	90710	101500
110BC02	22	110	200	38	101010	108850
120BC02	24	120	215	40	101010	110815
	6226	130	230	40	112780	120130
	28	140	250	42	127680	126510
	30	150	270	45	139745	135330
	32	160	290	48	151020	140235
	6234	170	310	52	184365	159850
	36	180	320	52	200060	169165
	38	190	340	55	235360	196130
	40	200	360	58	259880	205940

Table 2.2 (c) Deep groove ball bearings

ISI No.	Bearing No. (SKF)	d (mm)	D (mm)	B (mm)	Basic capacity, N	
					Static C_0	Dynamic C
10BC03	6300	10	35	11	3570	6080
12BC03	01	12	37	12	4220	7550
15BC03	02	15	42	13	5100	8580
17BC03	6303	17	47	14	6080	10300
20BC03	04	20	52	15	7550	12260
25BC03	05	25	62	17	10150	15985
30BC03	6306	30	72	19	14220	20990
35BC03	07	35	80	21	16970	25300
40BC03	08	40	90	23	20990	31380
45BC03	6309	45	100	25	29225	40700
50BC03	10	50	110	27	34720	47070
55BC03	11	55	120	29	41190	53940
60BC03	6312	60	130	31	47070	62270
65BC03	13	65	140	33	53450	71100
70BC03	14	70	150	35	60800	79925
75BC03	6315	75	160	37	71100	89280
80BC03	16	80	170	39	78450	94140
85BC03	17	85	180	41	85810	101500
90BC03	6318	90	190	43	96110	107870
95BC03	19	95	200	45	107870	117680
100BC03	20	100	215	47	129450	135330
105BC03	6321	105	225	49	140240	140235
110BC03	22	110	240	50	160730	152980
120BC03	24	120	260	55	162790	157890
	6326	130	280	58	191230	173580
	28	140	300	62	215750	196130
	30	150	320	65	249090	210840

Table 2.2 (d) Deep groove ball bearings

Bearing No. (SKF)	d (mm)	D (mm)	B (mm)	Basic capacity, N	
				Static C_0	Dynamic C
6403	17	62	17	12800	18000
6404	20	72	19	16600	24000
6405	25	80	21	20000	28250
6406	30	90	23	24000	33500
6407	35	100	25	32500	43000
6408	40	110	27	38000	50000
6409	45	120	29	46500	58500
6410	50	130	31	53000	70000
6411	55	140	33	64000	78500
6412	60	150	35	71000	84500
6413	65	160	37	80000	91500
6414	70	180	42	91000	100000
6415	75	190	45	101600	120000
6416	80	200	48	128000	130000
6417	85	210	52	138000	138000
6418	90	225	54	166000	152000

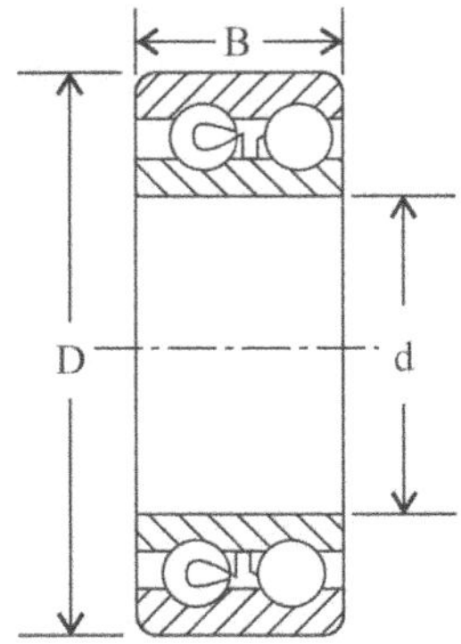

Table 2.2 (e) Self-aligning ball bearings

Bearing with cylindrical bore No. (SKF)	Bearing with taper bore No. (SKF)	d (mm)	D (mm)	B (mm)	Basic capacity, N	
					Static C_0	Dynamic C
2200	2204 K	10	30	14	1670	5540
01	05 K	12	32	14	1960	5640
02	2206 K	15	35	14	2110	5740
2203	07 K	17	40	16	2745	7500
04	08 K	20	47	18	3825	9610
05	2209 K	25	52	18	4120	9610
2206	10 K	30	62	20	5390	11770
07	11 K	35	72	23	7845	16180
08	2212 K	40	80	23	8825	16920
2209	13 K	45	85	23	9810	17410
10	–	50	90	23	10490	17410
11	2215 K	55	100	25	12455	20200
2212	16 K	60	110	28	15300	25810
13	17 K	65	120	31	19610	33340
14	2218 K	70	125	31	21080	34080
2215	19 K	75	130	31	21575	34080
16	20 K	80	140	33	24520	37850
17	–	85	150	36	29030	44620
2218	222 K	90	160	40	35550	53450
19		95	170	43	42170	63740
20		100	180	46	49030	76000
21		105	190	50	53940	82870
22		110	200	53	62270	96600

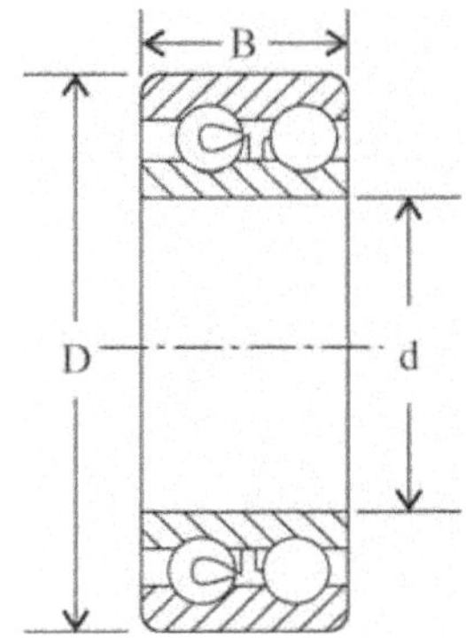

Table 2.2 (f) Self-aligning ball bearings

Bearing with cylindrical bore No. (SKF)	Bearing with taper bore No. (SKF)	d (mm)	D (mm)	B (mm)	Basic capacity, N	
					Static C_0	Dynamic C
2301	2304 K	12	37	17	2940	8925
02	05 K	15	42	18	3285	9070
03	06 K	17	47	19	4070	10790
2304	2307 K	20	52	21	5345	13730
05	08 K	25	62	24	7600	18490
06	09 K	30	72	27	9810	24030
2307	2310 K	35	80	31	12945	29810
08	11 K	40	90	33	15450	34720
09	12 K	45	100	36	19220	40940
2310	2313 K	50	110	40	23630	50750
11	-	55	120	43	28050	56390
12	15 K	60	130	46	32750	66693
2313	16 K	65	140	48	38490	73550
14	2317 K	70	150	51	43640	82375
15	18 K	75	160	55	50750	92670
16		80	170	58	56390	102970
2317		85	180	60	59580	106890
18		90	190	64	68160	115720

Table 2.2 (g) Single row angular contact ball bearings

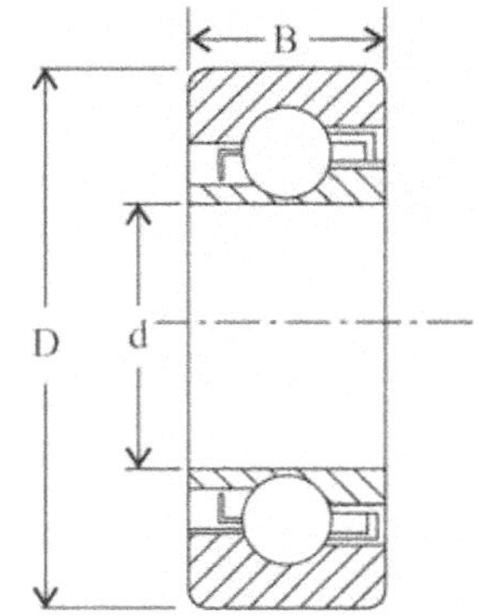

ISI No.	Bearing No. (SKF)	d (mm)	D (mm)	B (mm)	Basic capacity, N	
					Static C_0	Dynamic C
15BA02	7202 B	15	35	11	3680	6080
17BA02	03 B	17	40	12	4460	7700
20BA02	04 B	20	47	14	6370	10150
25BA02	7205 B	25	52	15	7700	11280
30BA02	06 B	30	62	16	10790	15400
35BA02	07 B	35	72	17	14710	20590
40BA02	7208 B	40	80	18	18490	24520
45BA02	09 B	45	85	19	21180	27655
50BA02	10 B	50	90	20	23140	28440
55BA02	7211B	55	100	21	29175	36285
60BA02	12 B	60	110	22	36285	42900
65BA02	13 B	65	120	23	42410	49030
70BA02	7214 B	70	125	24	44620	52470
75BA02	15 B	75	130	25	49030	54430
80BA02	16 B	80	140	26	55650	60800
85BA02	7217 B	85	150	28	63740	69630
90BA02	18 B	90	160	30	75760	81400
95BA02	19 B	95	170	32	85810	92670
100BA02	7220 B	100	180	34	90710	98070
105BA02	21 B	105	190	36	101500	106890
110BA02	22 B	110	220	38	112780	117680

Table 2.2 (h) Single row angular contact ball bearings

ISI No.	Bearing No. (SKF)	d (mm)	D (mm)	B (mm)	Basic capacity, N	
					Static C_0	Dynamic C
17BA03	7303 B	17	47	14	7110	11283
20BA03	04 B	20	52	15	8140	13580
25BA03	05 B	25	62	17	12260	18930
30BA03	7306 B	30	72	19	15840	24025
35BA03	07 B	35	80	21	20005	28050
40BA03	08 B	40	90	23	24910	34715
45BA03	7309 B	45	100	25	33340	44620
50BA03	10 B	50	110	27	40210	51485
55BA03	11 B	55	120	29	46580	59820
60BA03	7312 B	60	130	31	53450	69630
65BA03	13 B	65	140	33	61290	78450
70BA03	14 B	70	150	35	72570	87280
75BA03	7315 B	75	160	37	80415	96600
80BA03	16 B	80	170	39	89240	102970
85BA03	17 B	85	180	41	98070	111550
90BA03	7318 B	90	190	43	111500	120130
95BA03	19 B	95	200	45	122580	129250
100BA03	20 B	100	215	47	149060	144650
105BA03	7321 B	105	225	49	160090	153230
110BA03	22 B	110	240	50	188780	168920

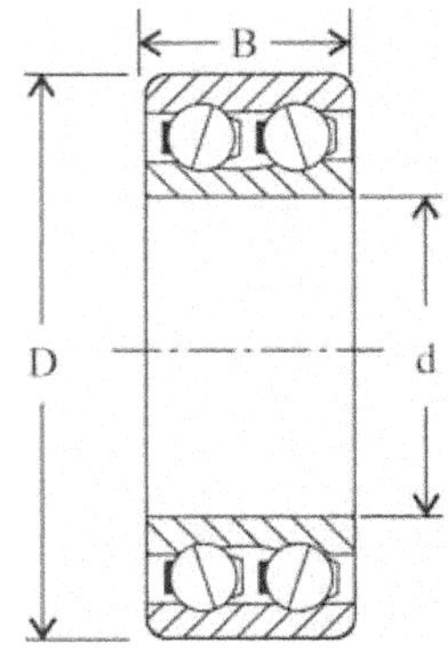

Table 2.2 (i) Double row angular contact ball bearings

Bearing No. (SKF)	d (mm)	D (mm)	B (mm)	Basic capacity, N	
				Static C_0	Dynamic C
3302 A	15	42	19.0	9070	13730
03 A	17	47	22.2	12650	18930
04 A	20	52	22.2	13730	18930
3305 A	25	62	25.4	19615	26085
06 A	30	72	30.2	27165	35300
07 A	35	80	34.9	35600	43640
3308 A	40	90	36.5	44620	53450
09 A	45	100	39.7	54430	62270
10 A	50	110	44.4	72570	80170
3311 A	55	120	49.2	78450	85910
12 A	60	130	54.0	94630	98070
13 A	65	140	58.7	108950	115720
3314 A	70	150	63.5	126510	135820
15 A	75	160	68.3	138080	140235
16 A	80	170	86.3	153965	157890
3317 A	85	180	73.0	173580	173580
18 A	90	190	73.0	205940	200150

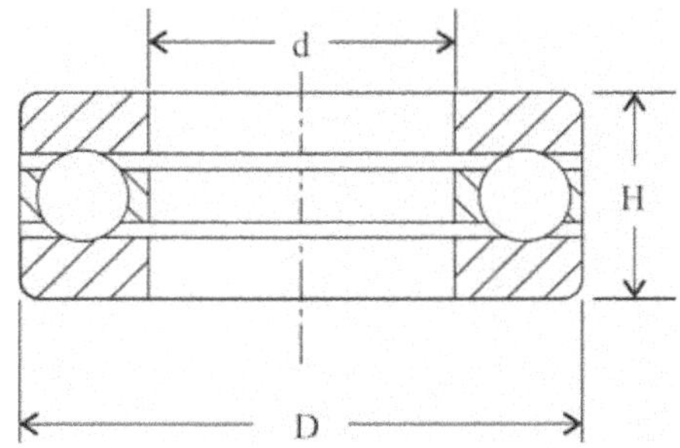

Table 2.2 (j) Single thrust ball bearings (with flat housing washer)

ISI No.	Bearing No.(SKF)	d (mm)	D (mm)	H (mm)	Basic capacity, N	
					Static C_0	Dynamic C
10TA12	51200	10	26	11	14000	10000
12TA12	01	12	28	11	15600	10400
15TA12	02	15	32	12	20400	12200
17TA12	51203	17	35	12	22000	12700
20TA12	04	20	40	14	31000	17300
25TA12	05	25	47	15	41500	21600
30TA12	51206	30	52	16	48000	22800
35TA12	07	35	62	18	64000	30500
40TA12	08	40	68	19	76500	34500
45TA12	51209	45	73	20	86500	36500
50TA12	10	50	78	22	91500	37500
55TA12	11	55	90	25	132000	55000
60TA12	51212	60	95	26	146000	57000
65TA12	13	65	100	27	156000	58500
70TA12	14	70	105	27	163000	60000
75TA12	51215	75	110	27	173000	61000
80TA12	16	80	115	28	180000	62000
85TA12	17	85	125	31	220000	75000
90TA12	51218	90	135	35	270000	91500
100TA12	20	100	150	38	340000	114000
110TA12	22	110	160	38	375000	120000
120TA12	51224	120	170	39	390000	120000
130TA12	26	130	190	45	510000	160000
140TA12	28	140	200	46	540000	163000
150TA12	51230	150	215	50	600000	176000
160TA12	32	160	225	51	620000	180000
170TA12	34	170	240	55	735000	212000
180TA12	51236	180	250	56	765000	216000
190TA12	38	190	270	62	915000	250000
200TA12	40	200	280	62	965000	255000
	51244	220	300	63	1040000	260000
	48	240	340	78	1430000	335000
	51252	260	360	79	1560000	345000
	56	280	380	80	1630000	355000
	60	300	420	95	2240000	455000
	51264	320	440	95	2360000	465000
	68	340	460	96	2450000	475000
	72	360	500	110	3150000	585000

Table 2.2 (k) Double thrust ball bearings (flat housing washers)

Bearing No. (SKF)	d (mm)	D (mm)	H (mm)	Basic capacity, N	
				Static C_0	Dynamic C
52202	15	32	22	20400	12200
04	20	40	26	31000	17300
05	25	47	28	41500	21600
52206	30	52	29	48000	22800
07	35	62	34	64000	30500
08	40	68	36	76500	34500
09	45	73	37	86500	36500
10	50	78	39	91500	37500
52211	55	90	45	132000	55000
12	60	95	46	146000	57000
13	65	100	47	156000	58500
14	70	105	47	163000	60000
15	75	110	47	173000	61000
52216	80	115	48	180000	62000
17	85	125	55	220000	75000
18	90	135	62	270000	91500
20	100	150	67	340000	114000

Table 2.2 (l) Cylindrical roller bearings

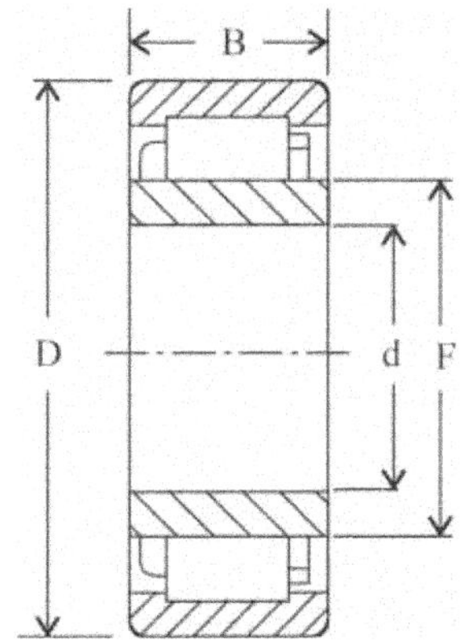

Bearing No. (SKF)	d (mm)	D (mm)	B (mm)	F (mm)	Basic capacity, N Static C_0	Dynamic C
NU 2205	25	52	18	32	11030	15790
2206	30	62	20	38.5	16970	23140
2207	51	72	23	43.8	27655	35600
NU 2208	40	80	23	50	32750	40700
2209	45	85	23	55	35600	43640
2210	50	90	23	60.4	38540	45500
NU 2211	55	100	25	66.5	45500	52560
2212	60	110	28	73.5	59820	69630
2213	65	120	31	79.6	74040	81640
NU 2214	70	125	31	84.5	78450	85910
2215	75	130	31	88.5	84730	96350
2216	80	140	33	95.3	98070	109340
NU 2217	85	150	36	101.8	117680	127000
2218	90	160	40	107	135820	140235
2219	95	170	43	113.5	160340	173580
NU 2220	100	180	46	120	184855	196130
2222	110	200	53	132.5	226435	254170
2224	120	215	58	143.5	271645	280470
NU 2226	130	230	64	156	306700	296650
2228	140	250	68	169	378540	362850
2230	150	270	73	182	436400	423650

Table 2.2 (m) Cylindrical roller bearings

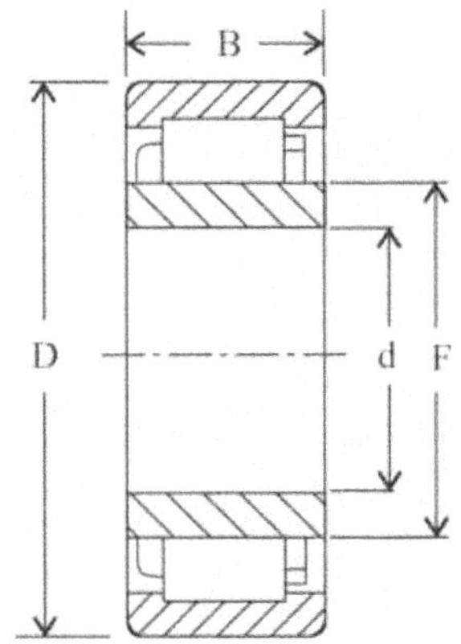

Bearing No. (SKF)	d (mm)	D (mm)	B (mm)	F (mm)	Basic capacity, N	
					Static C_0	Dynamic C
NU 2305	25	62	24	35	21280	30205
2306	30	72	27	42	26430	35305
2307	35	80	31	46.2	30695	41780
NU 2308	40	90	33	53.5	47270	58840
2309	45	100	36	58.5	55700	74040
2310	50	110	40	65	74040	90960
NU 2311	55	120	43	70.5	82965	106890
2312	60	130	46	77	103460	129250
2313	65	140	48	83.5	120130	142690
NU 2314	70	150	51	90	138270	155925
2315	75	160	55	95.5	166710	200150
2316	80	170	58	103	184855	213785
NU 2317	85	180	60	108	196130	227415
2318	90	190	64	115	222610	254000
2319	95	200	67	121.5	260760	292140
NU 2320	100	215	73	129.5	309890	341170
2322	110	240	80	143	423650	455030
2324	120	260	86	154	499160	557510
NU 2326	130	280	93	167	637430	609970
2328	140	300	102	180	726670	736480
2330	150	320	108	193	815910	815910

Table 2.2 (n) Taper roller bearings

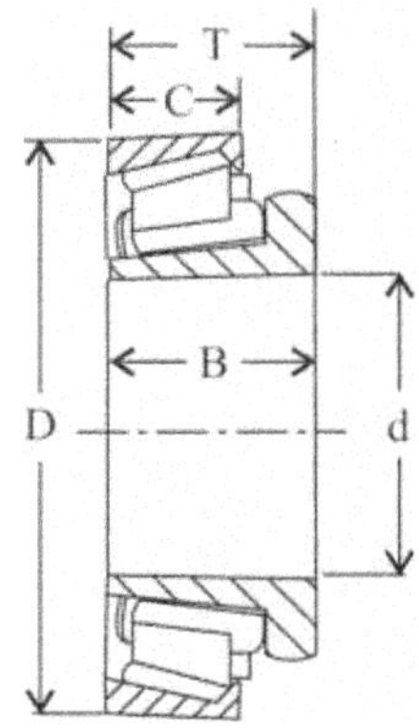

Bearing No. (SKF)	d (mm)	D (mm)	B (mm)	T (mm)	C (mm)	Basic capacity, N Static C_0	Dynamic C
32206	30	62	20	21.25	17	27165	31675
07	35	72	23	24.25	19	36285	41430
08	40	80	23	24.75	19	40210	45500
32209	45	85	23	24.75	19	45500	49915
10	50	90	23	24.75	19	47270	51580
11	55	100	25	26.75	21	61050	65115
32212	60	110	28	29.75	24	75710	78450
13	65	120	31	32.75	27	90810	96105
14	70	125	31	33.25	27	90810	96105
32215	75	130	31	33.25	27	99930	101450
16	80	140	33	35.25	28	113760	117680
17	85	150	36	38.5	30	134840	135820
32218	90	160	40	42.15	34	160730	158080
19	95	170	43	45.5	37	180440	180440
20	100	180	46	49	39	207705	202510
32221	105	190	50	53	43	241240	236340
22	110	200	53	56	46	271645	254000
24	120	213	53	61.5	50	334410	298120
32226	130	230	64	67.75	54	407960	369710
28	140	250	68	71.75	58	481510	422670
30	150	270	73	77	60	534460	480525

Table 2.2 (o) Spherical roller bearings

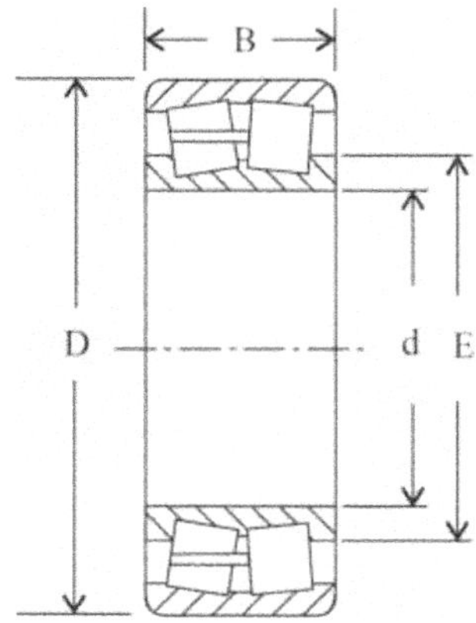

Bearing with cylindrical bore	d (mm)	D (mm)	B (mm)	E (mm)	Basic capacity, N	
					Static C_0	Dynamic C
*22205 C	25	52	18	32	22000	30000
*06 C	30	62	20	38	31000	45000
07 C	35	72	23	44	40500	52000
22208 C	40	80	23	50	48000	62000
09 C	45	85	23	55	52000	64000
10 C	50	90	23	60	55000	67000
22211 C	55	100	25	66	68000	83000
12 C	60	110	28	73	85000	100000
13 C	65	120	31	79	102000	118000
22214 C	70	125	31	84	108000	122000
15 C	75	130	31	90	112000	127000
16 C	80	140	33	95	140000	153000
22217 C	85	150	36	100	163000	180000
18 C	90	160	40	107	196000	208000
19 C	95	170	43	113	228000	245000
22220 C	100	180	46	120	255000	270000
22 C	110	200	53	132	345000	345000
24 C	120	215	58	143	400000	400000
22226 C	130	230	64	153	480000	465000
28 C	140	250	68	167	540000	530000
30 C	150	270	73	179	655000	640000
22232 C	160	290	80	191	780000	735000
34 C	170	310	86	204	900000	830000
36 C	180	320	86	213	950000	880000
22238 C	190	340	92	226	1020000	950000
40 C	200	360	98	238	1160000	1080000
44 C	220	400	108	264	1430000	1290000

Note: For bearings with taper bore add K-Example 22208 CK. Taper 1:12 on diameter.

*Not available with taper bore.

Table 2.2 (p) Needle bearings

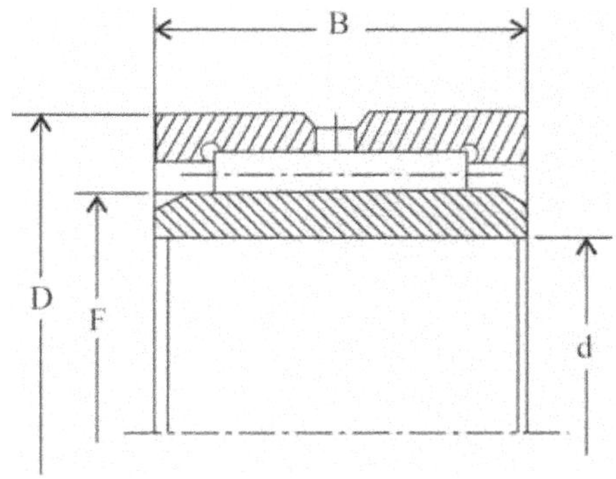

Bearing. No.	d (mm)	F (mm)	D (mm)	B (mm)	Basic capacity, N	
					Static C_0	Dynamic C
NA 49 02	15	20	28	13	7800	9400
03	17	22	30	13	8200	9700
04	20	25	37	17	14900	18500
49/22	22	28	39	17	16900	20100
05	25	30	42	17	17900	20800
49/28	28	32	45	17	18900	21500
06	30	35	47	17	19900	22100
49/32	32	40	52	20	26500	27000
07	35	42	55	20	27500	28000
08	40	48	62	22	37000	37500
09	45	52	68	22	39500	39000
49 10	50	58	72	22	42500	40500
12	60	68	85	25	57000	52000
14	70	80	100	30	84000	74000
16	80	90	110	30	94000	78000
18	90	105	125	35	134000	100000
20	100	115	140	40	144000	112000
24	120	135	165	45	210000	159000

Chapter 3

IC Engine Parts

3.1 I.C. Engine Cylinder

Bore diameter of two stroke cylinder, $D = \left[\dfrac{60 \times 10^6 \times 4 \times IP}{\pi p_m LN}\right]^{1/2}$ mm

Bore diameter of four stroke cylinder, $D = \left[\dfrac{60 \times 10^6 \times 8 \times IP}{\pi p_m LN}\right]^{1/2}$ mm

where

L – Length of stroke, mm
$= (1.2 \text{ to } 1.6)\ D$

N – Speed of the crank, rpm

p_m – Indicated mean effective pressure, N/mm^2

IP – Indicated power, kW
$= \dfrac{BP}{\eta_m}$

BP – Brake power, kW

η_m – Mechanical efficiency
$= 0.8 \text{ to } 0.9$

Thickness of cylinder, $t = \dfrac{pD}{2\sigma_t} + 6 \text{ to } 12$ mm $\qquad \left(\text{For} \dfrac{\sigma_t}{p} \geq 6\right)$

$= \dfrac{D}{2}\left[\left\{\dfrac{\sigma_t + p}{\sigma_t - p}\right\}^{1/2} - 1\right] + 8$ mm $\qquad \left(\text{For} \dfrac{\sigma_t}{p} < 6\right)$

p – Maximum gas pressure, N/mm^2

σ_t – Allowable tensile stress of cylinder material, N/mm^2

Thickness of cylinder head, $t_1 = 0.5\ D \sqrt{\dfrac{p}{\sigma_{th}}}$

Nominal diameter of bolt or stud, $d \simeq 1.2\ d_c$; $d_c = \sqrt{\dfrac{pD^2}{n\,\sigma_{tb}}}$

where n is the number of bolts connected with cylinder head.

$n = (0.1\ D \text{ to } 0.2\ D) + 4$

Length of cylinder, $L_c = L + L_p$ + Bottom clearance

L_p – Length of piston, mm

σ_t = 50 to 60 N/mm^2 for C.I

= 80 to 100 N/mm^2 for steel

σ_{tb} = 80 to 100 N/mm^2 (for bolt material)

σ_{th} = 30 to 50 N/mm^2 (for cylinder-head material)

Bottom clearance = 8 to 20 mm

3.2 Piston

Thickness of piston head based on the strength of piston material, $t_h = \sqrt{\dfrac{3pD^2}{16\sigma_{tp}}}$ mm

where

D – Diameter of piston (bore diameter of cylinder, mm)

p – Maximum gas pressure, N/mm^2

σ_{tp} – Allowable tensile stress of piston material, N/mm^2

= 35 to 40 N/mm^2 for C.I

= 60 to 100 N/mm^2 for steel

= 50 to 90 N/mm^2 for aluminium alloy

Thickness of piston head due to heat dissipation, $t_h = \dfrac{1000\ H}{12.56\ k(T_c - T_e)}$ mm

Heat flowing through the head, H = 0.05 m × HCV× BP kW

m – Mass of the fuel used, kg per brake power per second

Higher calorific value of the fuel, HCV = 44×10^3 kJ/kg for diesel fuel

= 11×10^3 kJ/kg for petrol fuel

$$\text{Brake power, } BP = \frac{p_m LAn}{60 \times 1000 \times 1000} \text{kW}$$

where

p_m – Brake mean effective pressure, N/mm^2

L – Length of stroke, mm

A – Area of the top side of piston, mm^2

n – Number of power strokes per minute

k – Heat conductivity factor

= 46.6×10^{-3} kW/m/°C for C.I

= 51×10^{-3} kW/m/°C for steel

= 175×10^{-3} kW/m/°C for aluminium alloys

$(T_c - T_e)$ = 220 °C for C.I

= 75 °C for aluminium alloys

where

T_c – Temperature at the centre of piston head, °C

T_e – Temperature at the edge of piston head, °C

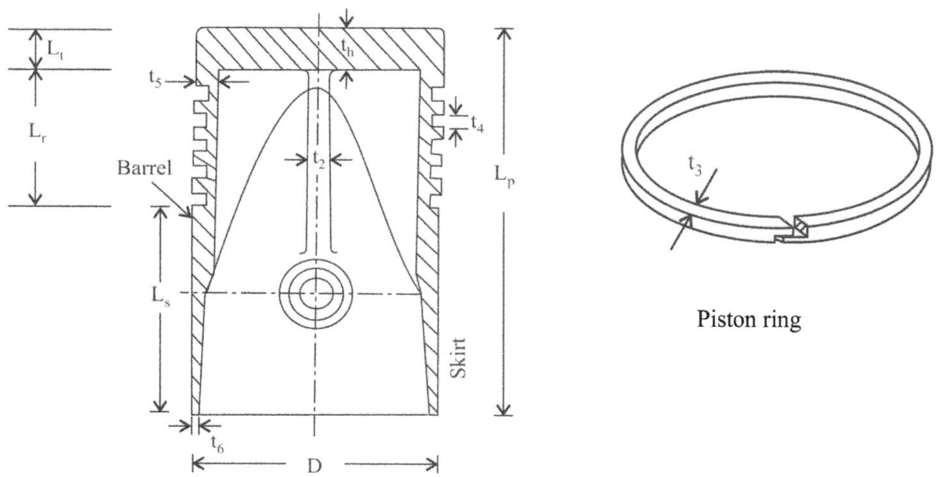

Fig. 3.1 IC engine piston

Thickness of rib, $t_2 = (0.3 \text{ to } 0.5)\ t_h$

Radial thickness of piston rings, $t_3 = D\sqrt{\dfrac{3p_c}{\sigma_{br}}}$

where σ_{br} – Allowable bending stress for ring material

$= 84$ to $112\ N/mm^2$ for alloy cast-iron

p_c – Contact pressure by the piston rings, usually 0.025 to 0.042 N/mm^2

Axial thickness of piston rings, $t_4 = (0.7 \text{ to } 1)\ t_3$

Minimum axial thickness of piston rings $= \dfrac{D}{10i}$

where

i – Number of rings

Radial depth of ring-groove, $b = t_3 + 0.4$ mm

Width of land between piston rings, $b_1 = (0.75 \text{ to } 1)\ t_4$

Thickness of barrel nearer to piston head, $t_5 = 0.03\ D + b + 4.5$ mm

Thickness of barrel at the open end of the piston, $t_6 = (0.25 \text{ to } 0.35)\ t_5$

Length of skirt, $L_S = (0.65 \text{ to } 0.80)\ D$

Length of ring section, $L_r = it_4 + (i - 1)\ b_1$

Width of top land, $L_t \simeq t_h$

Total length of piston, $L_p = L_s + L_r + L_t$

Diameter of Piston pin, $d_1 = \dfrac{F_g}{p_b \times l_1}$

where

p_b – Bearing pressure for piston pin, (12.5 to 15.4 N/mm^2)

Length of piston pin, $l_1 = 1.5\ d_1$

Maximum gas force, $F_g = \dfrac{\pi}{4} D^2 p$

Induced bending stress in piston pin, $\sigma_b = \dfrac{32\ M}{\pi d_1^3} < [\sigma_b]$

where bending moment, $M = \dfrac{F_g\ D}{8}$

$[\sigma_b] = 84$ N/mm^2 for case hardened steel

$= 140$ N/mm^2 for heat treated alloy steel

3.3 Connecting Rod

Load due to gas pressure, $F_g = \frac{\pi}{4} D^2 p$

where

D – Diameter of piston, mm

p – Pressure of gas, N/mm^2

Inertia force due to reciprocating parts, $F_i = \frac{W_r \omega^2 r}{g \times 1000}\left[\cos\theta + \frac{\cos 2\theta}{(l/r)}\right]$

where

W_r – Weight of reciprocating parts, N

ω – Angular speed of crank, rad/s

r – Radius of crank, mm

l – Actual length of connecting rod, mm

g – Acceleration due to gravity, m/s^2

θ – Angle of inclination of crank from inner dead centre, deg

Crippling load for connecting rod,

$$F_{cr} = \frac{\sigma_c A}{1 + a\left[\frac{l}{k_{xx}}\right]^2}$$

where

σ_c – Compressive yield stress, N/mm^2

(yield stress in compression can be about 330 N/mm^2)

A – Area of cross-section of connecting rod, mm^2

k_{xx} – Least radius of gyration, mm

$a = \frac{1}{7500}$ (Rankine's constant)

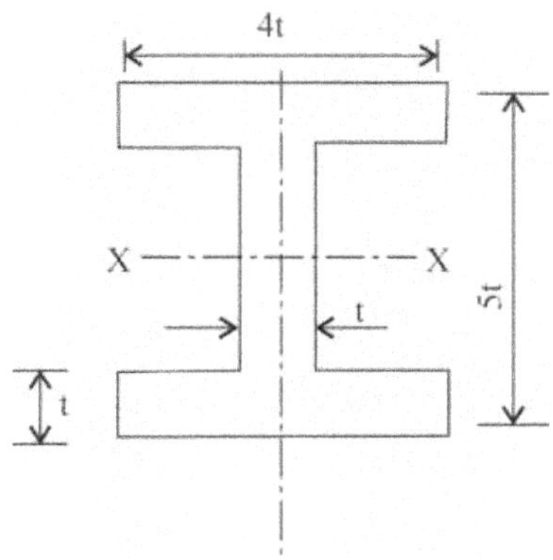

Fig. 3.2 I-section cross section of connecting rod

For I-section, $A = 11t^2$

$$I_{xx} = \frac{419t^4}{12} \qquad (I_{xx} - \text{Moment of inertia about xx – axis})$$

$$k_{xx}^2 = 3.18t^2$$

Axial load acting on the connecting rod, $F_a = \dfrac{(F_g + F_i)}{\sqrt{1-\left(\dfrac{\sin\theta}{l/r}\right)^2}} \simeq F_g$

(F_i and θ are very small values)

Induced maximum bending stress on the connecting rod, $\sigma_b = \dfrac{M_{max}}{Z}$

$$M_{max} = A\,\rho\,\omega^2\,r\,l^2/(9\sqrt{3}\times 10^{12})\,\text{N-mm}$$

where

ρ – Density of connecting rod material, kg/m^3

Z – Section modulus, mm^3

Maximum inertia force acting on bolts, $F_{im} = \dfrac{W_r}{g}\times\dfrac{\omega^2 r}{1000}\left[1+\dfrac{1}{(l/r)}\right]$

F_{im} is F_i at $\theta = 0$

Root diameter of bolt, $d_c = \left(\dfrac{2F_{im}}{\pi\sigma_t}\right)^{1/2}$

where

σ_t – Allowable tensile stress of bolt (120 to 175 N/mm^2)

Nominal or major diameter of the bolt, $d_b = 1.2\,d_c$

Number of bolts on the big end = 2

Diameter of piston-pin, $d_1 = \dfrac{F_g}{p_{b1} \times l_1}$

Diameter of crank-pin, $d_2 = \dfrac{F_g}{p_{b2} \times l_2}$

Bearing pressure for piston pin, p_{b1} = 12.5 to 15.4 N/mm^2

Bearing pressure for crank pin, p_{b2} = 10.8 to 12.6 N/mm^2

Length of piston pin, l_1 = (1.5 to 2) d_1

Length of crank pin, l_2 = (1.0 to 1.25) d_2

Thickness of bush, t_b = 2 to 5 mm

Thickness of big-end cap, $t_c = \left(\dfrac{F_{im} \times x}{b \times \sigma_{bc}} \right)^{1/2}$

where

σ_{bc} – Allowable bending stress of cap of big end (about 120 N/mm^2)

Distance between bolt centres in the big end, $x = d_2 + 2\,t_b + d_b + \text{clearance (3 mm)}$

Width of cap in big end of connecting rod, $b = l_2 - 2t_b$

3.4 Crankshaft

Symbols

D – Diameter of main journal, mm

d – Diameter of crank pin, mm

r – Radius of Crank, mm

F – Force transmitted from connecting rod, N

F_r – Radial component of force, N

F_t – Tangential component of force, N

L – Length of main journal, mm

l – Length of crank pin, mm

M – Bending moment, N-mm

p_b – Induced bearing pressure, N/mm^2

$[p_b]$ – Allowable bearing pressure, N/mm^2

σ – Overall stress induced in the web, N/mm^2

σ_a – Axial stress (i.e., tensile or compressive stress), N/mm^2

σ_b – Induced bending stress, N/mm^2

$[\sigma_b]$ – Allowable bending stress, N/mm^2

σ_{be} – Maximum equivalent bending stress, N/mm^2

σ_{br} – Bending stress by radial force, N/mm^2

σ_{bt} – Bending stress by tangential force, N/mm^2

τ – Induced shear stress, N/mm^2

$[\tau]$ – Allowable shear stress, N/mm^2

τ_e – Maximum equivalent shear stress, N/mm^2

T – Twisting moment, N-mm

t – Thickness of crank web, mm

W – Weight of flywheel, N

w – Width (i.e., mean width) of crank web, mm

x – Distance between the centre of crank pin and journal for overhung crankshaft (or) distance between the centres of main journals for centre crankshaft, mm

y – Distance of flywheel from main journal, mm

θ – Angle of inclination of the crank from inner dead centre, deg

ϕ – Angle of inclination of the connecting rod with the line of stroke, deg

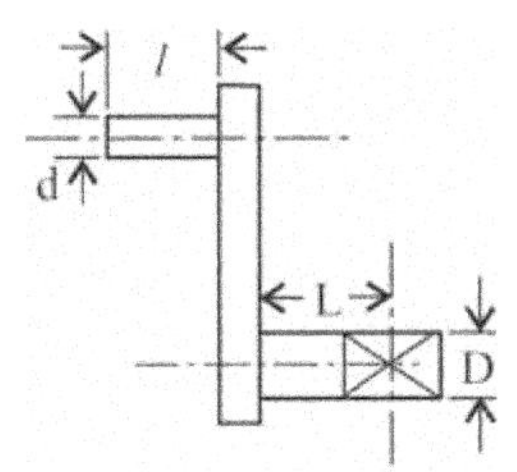

Fig. 3.3 Overhung crank

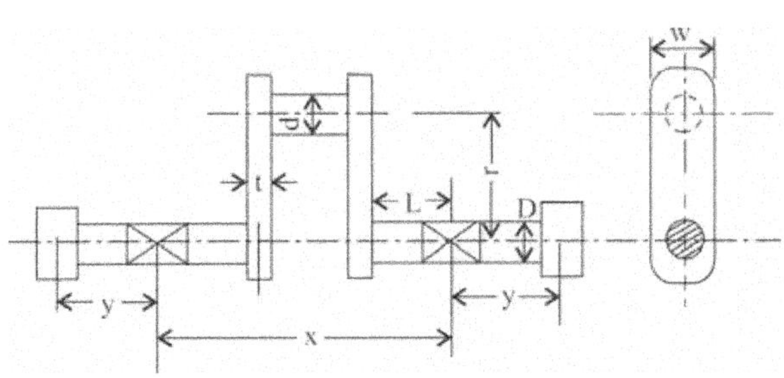

Fig. 3.4 Centre crank

Table 3.1 Stresses in crankshaft

Parts of crank shaft assembly and their induced stresses	Overhung crank	Centre crank
Crank pin:		
Direct shear stress	$\tau = \frac{4F}{\pi d^2} \le [\tau]$	$\tau = \frac{4F}{\pi d^2} \le [\tau]$
Bending stress	$\sigma_b = \frac{16Fl}{\pi d^3} \le [\sigma_b]$	$\sigma_b = \frac{8\,Fx}{\pi d^3} \le [\sigma_b]$
Bearing pressure	$p_b = \frac{F}{l\,d} \le [p_b]$	$p_b = \frac{F}{l\,d} \le [p_b]$
Crank-web:		
Radial component of force	$F_r = F \cos(\theta + \phi)$	$F_r = \frac{F}{2}\cos(\theta + \phi)$
Tangential component of foce	$F_t = F \sin(\theta + \phi)$	$F_t = \frac{F}{2}\sin(\theta + \phi)$
Axial stress by radial force	$\sigma_a = \frac{F_r}{wt} \le [\sigma_b]$	$\sigma_a = \frac{F_r}{wt} \le [\sigma_b]$
Bending stress by radial force	$\sigma_{br} = \frac{3\,F_r(l + t)}{wt^2} \le [\sigma_b]$	$\sigma_{br} = \frac{3F_r[x - (l + t)]}{wt^2} \le [\sigma_b]$
Bending stress by tangential force	$\sigma_{bt} = \frac{6\,F_t r}{t\,w^2} \le [\sigma_b]$	$\sigma_{bt} = \frac{6\,F_t\,r}{tw^2} \le [\sigma_b]$
Overall stress acting on the web	$\sigma = \sigma_a + \sigma_{br} + \sigma_{bt} \le [\sigma_b]$	$\sigma = \sigma_a + \sigma_{br} + \sigma_{bt} \le [\sigma_b]$
Crank-shaft journal:		
Bending stress by radial force	$\sigma_{br} = \frac{32\,F_r\,x}{\pi D^3} \le [\sigma_b]$	–
Bending moment	$M = F_t x$	$M = Wy$
Twisting moment	$T = F_t r$	$T = F_t r$
Maximum equivalent bending stress	$\sigma_{be} = \frac{16(M + \sqrt{M^2 + T^2})}{\pi D^3} \le [\sigma_b]$	$\sigma_{be} = \frac{16(M + \sqrt{M^2 + T^2})}{\pi D^3} \le [\sigma_b]$
Maximum equivalent shear stress	$\tau_e = \frac{16\sqrt{M^2 + T^2}}{\pi D^3} \le [\tau]$	$\tau_e = \frac{16\sqrt{M^2 + T^2}}{\pi D^3} \le [\tau]$
Bearing pressure	$p_b = \frac{F}{LD} \le [p_b]$	$p_b = \frac{F}{2\,LD} \le [p_b]$

Table 3.2 Usual proportions

Overhung crank	Centre crank
D = (1.25 to 1.5) d	D = d
L = 1.25 D	L = 1.25 D
l = (1 to 1.25) d	l = (1 to 1.25) d
t = (0.7 to 1.0) d	t = (0.7 to 1.0) d
$w = \frac{1.5(d+D)}{2}$	w = 1.5 d

Allowable stresses for mild steel crankshaft:

$[\sigma_b]$ = 60 to 100 N/mm^2

$[\tau]$ = 40 to 60 N/mm^2

Allowable bearing pressure in crank pin, $[p_b]$ = 4 to 12 N/mm^2

Allowable bearing pressure in main shaft, $[p_b]$ = 1.5 to 12 N/mm^2

Chapter 4

Power Transmission Systems, Pulleys

4.1 Flat Belts

Symbols

C – Centre distance between pulleys, mm
D – Diameter of bigger pulley, mm
d – Diameter of smaller pulley, mm
g – Acceleration due to gravity, m/s^2
L – Length of belt, mm
N – Speed of bigger pulley, rpm
n – Speed of smaller pulley, rpm
T_1 – Tension in tight side of belt, N
T_2 – Tension in slack side of belt, N
T_{t_1} – Total tension in tight side, N
T_{t_2} – Total tension in slack side, N
T_c – Centrifugal force, N
t – Thickness of belt, mm
w – Weight of the belt per metre length, N/m
θ – Angle (or Arc) of contact of the belt with the smaller pulley, deg
μ – Coefficient of friction between the contact surfaces of belt and pulley
σ_t – Allowable tensile stress for the belt material, N/mm^2

Power transmitted by belt, $P = \dfrac{(T_1 - T_2)v}{1000}$ kW

Belt velocity, $v = \frac{\pi dn}{60000} = \frac{\pi DN}{60000}$ m/s

Ratio of tensions of the belt, $\frac{T_1}{T_2} = e^{\mu\theta'}$

[neglecting centrifugal force i.e., for low velocity]

$\frac{T_{t_1} - T_c}{T_{t_2} - T_c} = e^{\mu\theta'}$ [considering centrifugal force i.e., for high velocity]

$T_c = \frac{w}{g} v^2;$ $\theta' = \frac{\pi\theta}{180}$ radians

Maximum tension in the belt, $T_{t_1} = \sigma_t \times b \times t$

Width of belt in mm, $b = \frac{\text{Maximum tension}}{\text{Allowable tension per mm width}}$

Type of belt drive	Arc of contact (θ), deg.	Length of belt (L), mm
Open belt drive	$\theta = 180^\circ - \left(\frac{D-d}{C}\right)60^\circ$	$L = 2C + \frac{\pi}{2}(D+d) + \frac{(D-d)^2}{4C}$
Cross belt drive	$\theta = 180^\circ + \left(\frac{D+d}{C}\right)60^\circ$	$L = 2C + \frac{\pi}{2}(D+d) + \frac{(D+d)^2}{4C}$

4.2 Pulleys for Flat Belts

Pulley Rim

Centrifugal stress induced in the rim of the pulley, $\sigma = \rho V^2$ N/m^2

ρ – Density of the rim material, 7200 kg/m^3 for cast iron

V – Velocity of the rim, m/s

Width of the pulley, B = 1.25 b

where b – Width of belt

Thickness of pulley rim in mm, $t = \frac{D}{300} + 2\,\text{mm}$ for single belt

$= \frac{D}{200} + 6\,\text{mm}$ for double belt

where

D is diameter of the pulley in mm.

Dimensions of Arms

The number of arms, n = 4 for pulley diameter from 200 mm to 600 mm

= 6 for pulley diameter from 600 mm to 1500 mm

The cross-section of the arms is elliptical with major axis (2a) equal to twice the minor axis (2b).

Maximum bending moment on the arm at the hub end, $M = \frac{2T}{n}$

where T is the torque transmitted.

Section modulus, $Z = \frac{\pi}{4} a^2 b$

Bending stress, $\sigma_b = \frac{M}{Z}$

Hub

The diameter of the hub in mm, $d_1 = 1.5\,d + 25$ mm

where

d is shaft diameter in mm.

d_1 should not be greater than 2d.

The length of the hub, $L = \frac{\pi}{2} \times d$

The minimum length of hub is $\frac{2}{3}B$, but it should not be more than B, the width of pulley.

4.3 V-Belt Drives

Symbols

D – Diameter (i.e., pitch diameter) of bigger pulley, mm

d – Diameter (i.e., pitch diameter) of smaller pulley, mm

K_a – Arc of contact factor

K_l – Pitch length (of belt) factor

K_s – Service factor

N – Speed of bigger pulley in rpm

n – Speed of smaller pulley in rpm

P_r – Rated (i.e., specified) power, kW

P – Belt rating (i.e., power transmitting capacity) of single belt, kW

T_1 – Tension in tight side of belt, N

T_2 – Tension in slack side of belt, N

α – Angle subtended by sides of V-belt, deg

θ – Angle (or Arc) of contact of belt with the smaller pulley, deg

μ – Coefficient of friction between the contact surfaces of belt and pulley

Power transmitted by belt, $P' = \dfrac{(T_1 - T_2)v}{1000}$ kW

Belt velocity, $v = \dfrac{\pi\, dn}{60000} = \dfrac{\pi\, DN}{60000}$ m/s

Ratio of tensions of the belt, $\dfrac{T_1}{T_2} = e^{\mu\, \theta'/\sin(\alpha/2)}$; $\theta' = \dfrac{\pi\, \theta}{180}$ radians

Design power, $P_d = \dfrac{P_r \times K_s}{K_a \times K_l}$

Number of belts, $i = \dfrac{P_d}{P}$

[For P (i.e., belt rating) refer table 4.4]

Length of belt, $L = 2C + \dfrac{\pi}{2}(D + d) + \dfrac{(D - d)^2}{4C}$

Centre distance between pulleys, $C = A + \sqrt{A^2 - B}$

where $A = \dfrac{L}{4} - \pi\left(\dfrac{D + d}{8}\right)$ and $B = \dfrac{(D - d)^2}{8}$

Maximum centre distance, $C_{max} = 2(D + d)$

Table 4.1 Correction factors for industrial service, K_s

Severity of service	Type of driven machines	Types of driving units					
		AC motors; Normal torque. Squirrel cage, Synchronous and split phase DC motors; Shunt wound multiple cylinder Internal combustion engines over 10 rev/s			AC motors; High torque, High slip repulsion induction, Single phase series wound and slip ring DC motors; Series wound and compound wound, Single cylinder internal combustion engines. Multiple cylinder internal combustion engines under 10 rev/s, Line shafts, Clutches, Brakes, Direct on Line starting.		
		Upto 10 hr	Over 10 hr to 16 hr	Over 16 hr and continu-ous service	Upto 10 hr	Over 10 hr to 16 hr	Over 16 hr and continu-ous service
Light-duty	Agitators for liquids, blowers and exhausters, centrifugal pumps and compressors, fans upto 7.5 kW and light-duty conveyors	1.0	1.1	1.2	1.1	1.2	1.3
Medium duty	Belt conveyors for sand, grain etc; dough mixers; fans over 7.5 kW; generators; line shafts; laundry machinery; machine tools; punches, presses and shears; printing machinery; positive displacement rotary pumps; and revolving and vibrating screens.	1.1	1.2	1.3	1.2	1.3	1.4
Heavy duty	Brick machinery, bucket elevators, exciters, piston compressors, conveyors (drag-pan-screw), hammer mills, paper mill beaters, piston pumps, positive displacement blowers, pulverizers, sawmill and wood-working machinery, and textile machinery.	1.2	1.3	1.4	1.4	1.5	1.6
Extra-heavy duty	Crushers (gyratory-jaw-roll), mills (ball-rod-tube), hoists, and rubber (calenders-extruders-mills)	1.3	1.4	1.5	1.5	1.6	1.8

Table 4.2 Correction factors for arc of contact, K_a

$$\text{Arc of contact angle} = 180^\circ - \left(\frac{D-d}{C}\right)60^\circ$$

Arc of contact on smaller pulley (in degrees)	Correction factor (Proportion of 180° rating)	
	V-V	V-Flat
180	1.00	0.75
177	0.99	0.76
174	0.99	0.76
171	0.98	0.77
169	0.97	0.78
166	0.97	0.79
163	0.96	0.79
160	0.95	0.80
157	0.94	0.81
154	0.93	0.81
151	0.93	0.82
148	0.92	0.83
145	0.91	0.83
142	0.90	0.84
139	0.89	0.85
136	0.88	0.85
133	0.87	0.86
130	0.86	0.86
127	0.85	0.85
123	0.83	0.83
120	0.82	0.82
117	0.81	0.81
113	0.80	0.80
110	0.78	0.78
106	0.77	0.77
103	0.75	0.75
99	0.73	0.73
95	0.72	0.72
91	0.70	0.70
87	0.68	0.68
83	0.65	0.65

Table 4.3 Nominal inside length, nominal pitch length and correction factors

Nominal pitch length						**Pitch length variation**		**Correction factor (K_l)**				
Nominal inside length	**Cross-section A**	**Cross-section B**	**Cross-section C**	**Cross-section D**	**Cross-section E**	**Pitch-length limits**	**Maximum variation in length within a matched set**	**Belt cross-section symbols**				
(mm)	**(mm)**	**(mm)**	**(mm)**	**(mm)**	**(mm)**	**(mm)**	**(mm)**	**A**	**B**	**C**	**D**	**E**
965	1001	1008	–	–	–			0.88	0.83	–	–	–
991	1026	–	–	–	–			0.88	–	–	–	–
1016	1051	1059	–	–	–	+ 14.0	2.5	0.89	0.84	–	–	–
1067	1102	1110	–	–	–	– 8.9		0.90	0.85	–	–	–
1092	1128	–	–	–	–			0.90	–	–	–	–
1168	1204	1212	–	–	–			0.92	0.87	–	–	–
1219	1255	1262	–	–	–			0.93	0.88	–	–	–
1295	1331	1339	1351	–	–			0.94	0.89	0.80	–	–
1372	–	1415	–	–	–			–	0.90	–	–	–
1397	1433	1440	–	–	–	+ 16.0		0.96	0.90	–	–	–
1422	1458	1466	–	–	–	– 9.0		0.96	0.90	–	–	–
1473	1509	–	–	–	–			0.97	–	–	–	–
1524	1560	1567	1580	–	–		5.0	0.98	0.92	0.82	–	–
1600	1636	–	–	–	–			0.99	–	–	–	–
1626	1661	–	–	–	–	+ 17.8		0.99	–	–	–	–
1651	1687	1694	–	–	–	– 12.5		1.00	0.94	–	–	–
1727	1763	1770	1783	–	–			1.00	0.95	–	–	–
1778	1814	1821	–	–	–			1.01	0.95	0.85	–	–
1905	1941	1948	1961	–	–	+ 17.8 – 12.5		1.02	0.97	0.87	–	–
1981	2017	2024	–	–	–			1.03	0.98	–	–	–
2032	2068	–	–	–	–		7.5	1.04	–	–	–	–
2057	2093	2101	2113	–	–	+ 30		1.04	0.98	0.89	–	–
2159	2195	2202	2215	–	–	– 16		1.05	0.99	0.90	–	–
2286	2322	2329	2342	–	–			1.06	1.00	0.91	–	–
2438	2474	–	2494	–	–			1.08	–	0.92	–	–
2464	–	2507	–	–	–			–	1.02	–	–	–
2540	–	2583	–	–	–			–	1.03	–	–	–
2667	2703	2710	2723	–	–	+ 34		1.10	1.04	0.94	–	–
2845	2880	2888	2901	–	–	– 18		1.11	1.05	0.95	–	–
3048	3084	3091	3104	3127	–			1.13	1.07	0.97	0.86	–
3150	–	–	3205	–	–		10	–	–	0.97	–	–
3251	3287	3294	3 307	3330	–	+ 38 –21		1.14	1.08	0.98	0.87	–
3404	–	–	3 459	–	–			–	–	0.99	–	–
3658	3693	3701	3 713	3736	–			–	1.11	1.00	0.90	–
4013	–	4056	4069	4092	–	+43 – 24		–	1.13	1.02	0.92	–

Table 4.3 *Contd...*

4115	–	4158	4171	4194	–			–	1.14	1.03	0.92	–
4394	–	4437	4450	4473	–	+ 43		–	1.15	1.04	0.93	–
4572	–	4615	4628	4651	–	– 24		–	1.16	1.05	0.94	–
4953	–	4996	5009	5032	–	+ 49	12.5	–	1.18	1.07	0.96	–
5334	–	5377	5390	5413	5426	– 28		–	1.19	1.08	0.96	0.94
6045	–	–	6101	6124	6137			–	–	1.11	1.00	0.96
6807	–	–	6863	6886	6899	+ 56 – 32		–	–	1.14	1.03	0.99
7569	–	–	7625	7648	7661	+ 56 – 32	15	–	–	1.16	1.05	1.01
8331	–	–	8387	8410	8423	+ 65		–	–	1.19	1.07	1.03
9093	–	–	9149	9172	9185	– 37		–	–	1.21	1.09	1.05
9855	–	–	–	9934	9947	+ 65 – 37		–	–	1.23	1.11	1.07
10617	–	–	–	10696	10709	+76		–	–	1.24	1.12	1.09
12141	–	–	–	12220	12233	– 43	17.5	–	–	–	1.16	1.12
13665	–	–	–	13744	13757	+ 89 – 50		–	–	–	1.18	1.14
15189	–	–	–	15268	15281	+ 105		–	–	–	1.20	1.17
16713	–	–	–	16792	16805	– 59		–	–	–	1.23	1.19

Table 4.4 Formulae for power transmitting capacities of single belt

Usual load of drive, kW	Belt type	Belt rating (kW) formula	Maximum value of d_e in the formula mm
0.75 – 5	A	$P = \left(0.45\,v^{-0.09} - \frac{19.62}{d_e} - 0.765\times10^{-4}v^2\right)v$	125
2 – 15	B	$P = \left(0.79\,v^{-0.09} - \frac{50.8}{d_e} - 1.32\times10^{-4}v^2\right)v$	175
7.5 – 75	C	$P = \left(1.47\,v^{-0.09} - \frac{142.7}{d_e} - 2.34\times10^{-4}v^2\right)v$	300
22 – 150	D	$P = \left(3.22\,v^{-0.09} - \frac{506.7}{d_e} - 4.78\times10^{-4}v^2\right)v$	425
30 –190	E	$P = \left(4.58\,v^{-0.09} - \frac{952}{d_e} - 7.05\times10^{-4}v^2\right)v$	700

where

P – Maximum power in kW at 180° arc of contact for a belt of average length

d_e – Equivalent pitch diameter = $d \times F_d$

F_d – Small diameter factor to account for variation of arc of contact

Table 4.5 Small diameter factors, F_d

Speed ratio range, D/d	F_d	Speed ratio range, D/d	F_d
1.0 to 1.019	1.00	1.223 to 1.274	1.08
1.02 to 1.032	1.01	1.275 to 1.340	1.09
1.033 to 1.055	1.02	1.341 to 1.429	1.10
1.056 to 1.081	1.03	1.43 to 1.562	1.11
1.082 to 1.109	1.04	1.563 to 1.814	1.12
1.11 to 1.142	1.05	1.815 to 2.948	1.13
1.143 to 1.178	1.06	2.949 and over	1.14
1.179 to 1.222	1.07		

4.4 Chain Drives

Symbols

A – Projected bearing area, mm^2

C – Centre distance between sprockets, mm

C_p – Centre distance in multiples of pitches

d_1 – Pitch diameter of pinion-sprocket, mm

d_2 – Pitch diameter of wheel-sprocket, mm

F – Load due to design power, N

F_c – Centrifugal tension, N

F_s – Tension due to sagging, N

F_t – Tangential force due to power transmission, N

g – Acceleration due to gravity, m/s^2

i – Transmission ratio or speed ratio

k – Coefficient for sag

N – Assumed factor of safety related with speed

[N] – Actual factor of safety

F_{ser} – Service factor

F_1 – Load factor

F_2 – Factor for distance regulation

F_3 – Factor for centre distance of sprockets

F_4 – Factor for position of sprockets

F_5 – Lubrication factor

F_6 – Rating factor

l – Length of chain, mm

l_p – Length of chain in multiples of pitches

n_1 – Speed of pinion-sprocket, rpm

n_2 – Speed of wheel-sprocket, rpm

P_d – Design power, kW

P_r – Rated power, kW

p – Pitch of the chain, mm

Q – Maximum operating load, N

[Q] – Breaking load for the chain, N

v – Velocity of chain, m/s

w – Weight of chain per metre length, N/m

Z_1 – Number of teeth of pinion sprocket

Z_2 – Number of teeth of wheel sprocket

σ – Induced bearing stress, N/mm^2

$[\sigma]$ – Allowable bearing stress, N/mm^2

ΔC – Centre distance decrement, mm

Design power, $P_d = P_r \times F_{ser}$

$$F_{ser} = F_1 \times F_2 \times F_3 \times F_4 \times F_5 \times F_6 \quad \text{(Refer Table 4.6)}$$

Load applied in the chain due to design power, $F = \dfrac{P_d \times 1000}{v}$

Maximum operating load, $Q = F \times N < [Q]$

Induced bearing stress due to power transmission, $\sigma = \dfrac{F}{A} < [\sigma]$

Chain velocity, $v = \dfrac{Z_1 n_1 p}{60 \times 1000} = \dfrac{Z_2 n_2 p}{60 \times 1000}$

$$= \frac{\pi d_1 n_1}{60 \times 1000} = \frac{\pi d_2 n_2}{60 \times 1000}$$

The empirical formula to determine pitch in mm, $p \le 10\left[\dfrac{60.67}{n'}\right]^{2/3}$

where n' – speed of small sprocket, rps

$$= \frac{n_1}{60}$$

Pitch diameter of pinion-sprocket, $d_1 = \dfrac{p}{\sin\left(\dfrac{180^\circ}{Z_1}\right)}$

Pitch diameter of wheel-sprocket, $d_2 = \dfrac{p}{\sin\left(\dfrac{180^\circ}{Z_2}\right)}$

Transmission ratio, $i = \dfrac{Z_2}{Z_1} = \dfrac{n_1}{n_2}$

Actual factor of safety (for checking), $[N] = \dfrac{[Q]}{\Sigma F}$ where $\Sigma F = F_t + F_c + F_s$

$$F_t = \frac{P_r}{v};\ F_c = \frac{w\,v^2}{g};\ F_s = \frac{k \times w \times C}{1000}$$

Approximate centre distance in multiples of pitches, $C_p = \dfrac{C_0}{p}$

where C_0 is initially assumed centre distance.

Length of chain in multiples of pitches (i.e., approximate number of links),

$$l_p = 2C_p + \frac{Z_1 + Z_2}{2} + \frac{\left(\dfrac{Z_2 - Z_1}{2\pi}\right)^2}{C_p}$$

(to be corrected to even number)

Actual length of chain, $l = l_p \times p$ mm

Final centre distance corrected to even number of pitches, $C = \left(\dfrac{e + \sqrt{e^2 - 8m}}{4}\right)p$

where $e = l_p - \dfrac{Z_1 + Z_2}{2}$

$m = \left(\dfrac{Z_2 - Z_1}{2\pi}\right)^2$, constant (Refer Table 4.11)

Centre distance decrement (i.e., allowance) to provide initial sag, $\Delta C = 1\,\%$ of C (i.e., 0.01 C)

Table 4.6 Factors

Load factor	F_1
Constant load	1.0
Variable load or load with mild shocks	1.25
Variable load or load with heavy shocks	1.5
Factor for distance regulation	F_2
Adjustable supports	1.0
Drive using idler sprocket	1.1
Fixed centre distance	1.25
Factor for centre distance of sprockets	F_3
$\frac{l_p}{Z_1 + Z_2} > 1$ or $C_p < 25p$	1.25
$\frac{l_p}{Z_1 + Z_2} = 1.5$ or $C_p = 30$ to $50\ p$	1
$\frac{l_p}{Z_1 + Z_2} \geq 2.0$ or $C_p = 60$ to $80\ p$	0.8
Factor for the position of sprockets	F_4
Inclination of the line joining the centres of the sprockets to the horizontal upto 60°	1
More than 60°	1.25
Lubrication factor	F_5
Continuous (oil-bath or forced lubrication)	0.8
Drop-lubrication	1.0
Periodic	1.5
Rating factor	F_6
Single shift of 8 hours a day	1.0
Double shift of 16 hours a day	1.25
Continuous running	1.5

Preferred values of transmission ratio, i

1, 1.12, 1.25, 1.4, 1.6, 1.8, 2, 2.25, 3.15, 4, 4.5, 5, 5.6, 6.3, 7.1

Table 4.7 Recommended number of teeth on sprocket

Transmission ratio, i	1-2	2-3	3-4	4-5	5-7
Number of teeth on sprocket, Z_1	30-27	27-25	25-23	23-21	21-17

Where space is problem, Z_1 minimum = 7

Number of teeth on sprocket wheel, $Z_2 = i\ Z_1$

$$Z_{2\ max} = 100 \text{ to } 120$$

Note: When Z_2 is very large, chain slips off the sprocket for a small pull.

Table 4.8 Minimum value of factor of safety, N (for $F_{ser} = 1$ & $Z_1 = 15$ to 30)

n_1	< 50	200	400	600	800	1000	1200	1600	2000	2400	2800
N	7.0	7.8	8.55	9.35	10.2	11.0	11.7	13.2	14.8	16.3	18.0

Table 4.9 Allowable bearing stress [σ], N/mm² (for $F_{ser} = 1$ & $Z_1 = 15$ to 30)

n_1	< 50	200	400	600	800	1000	1200	1600	2000	2400	2800
[σ]	34	31	28	26	24	22	21	18	16	15	14

Table 4.10 Coefficient for sag, k

Coefficient for sag	**Position of chain drive**			
	Horizontal	**Upto 40°**	**More than 40°**	**Vertical**
k	6	4	2	1

Table 4.11 Value of $m = \left(\frac{Z_2 - Z_1}{2\pi}\right)^2$

$Z_2 - Z_1$	m	$Z_2 - Z_1$	m	$Z_2 - Z_1$	m	$Z_2 - Z_1$	m
1	0.025	26	17.12	51	66.0	76	146.3
2	0.101	27	18.47	52	68.5	77	150.2
3	0.228	28	19.86	53	71.2	78	154.1
4	0.405	29	21.3	54	73.9	79	158.1
5	0.633	30	22.8	55	76.6	80	162.1
6	0.912	31	24.3	56	79.4	81	166.2
7	1.24	32	25.9	57	82.3	82	170.3
8	1.62	33	27.6	58	85.2	83	174.5
9	2.05	34	29.3	59	88.2	84	178.7
10	2.53	35	31.0	60	91.2	85	183.0
11	3.07	36	32.8	61	94.3	86	187.3
12	3.65	37	34.7	62	97.4	87	191.7
13	4.28	38	36.6	63	100.5	88	196.2
14	4.97	39	38.5	64	103.8	89	200.6
15	5.70	40	40.5	65	107.0	90	205.2
16	6.49	41	42.6	66	110.3	91	209.8
17	7.32	42	44.7	67	113.7	92	214.4
18	8.21	43	46.8	68	117.1	93	219.1
19	9.14	44	49.0	69	120.6	94	223.8
20	10.13	45	51.3	70	124.1	95	228.6
21	11.17	46	53.6	71	127.7	96	233.4
22	12.26	47	56.0	72	131.3	97	238.3
23	13.40	48	58.4	73	135.0	98	243.3
24	14.59	49	60.8	74	138.7	99	248.3
25	15.83	50	63.3	75	142.9	100	253.3

4.5 Fibre Ropes

Symbols

D_1 – Diameter of smaller pulley, mm

D_2 – Diameter of larger pulley, mm

x – Centre distance, mm

x_{min} – Minimum centre distance of rope drives, mm

2ϕ – Groove angle, deg

T_1 – Tension on the tight side, N

T_2 – Tension on the slack side, N

T_c – Centrifugal tension, N

μ – Coefficient of function

θ – Arc of contact on smaller pulley, rad

V – Velocity of rope, m/s

g – Acceleration due to gravity, m/s^2

Design Relations

For manila ropes, Ultimate tensile breaking load = 5000 d^2 Newtons where d is rope diameter in cm.

For cotton ropes, Ultimate tensile breaking load = 3500 d^2 Newtons where d is rope diameter in cm.

Average value for the weight of rope, w = 0.70 d^2 N/m where d is in cm.

Minimum value for factor of safety = 35 for manila ropes

= 30 for cotton ropes

Sheave diameter = 40d where d is the rope diameter in cm.

Length of the rope, $L = \frac{\pi}{2}(D_1 + D_2) + (D_2 - D_1)\sin^{-1}\frac{D_2 - D_1}{2x} + 2x\cos\phi$

$$x_{min} = (D_2 - D_1) + \frac{3}{2}D_2$$

$$\frac{T_1}{T_2} = e^{\mu\theta\ \mathrm{cosec}\phi}$$

If centrifugal tension is taken into consideration, $\frac{T_1 - T_c}{T_2 - T_c} = e^{\mu\theta\ \mathrm{cosec}\phi}$

where $$T_c = \frac{wV^2}{g}$$

$$\text{Total power transmitted in kW} = \frac{(T_1 - T_2)V \times n}{1000}$$

where n – Number of ropes

Fibrous ropes used in hoisting tackle

$$P = C \times F$$

P – Effort

F – Weight attached

C – Pulley factor of resistance

= 1.15 for poorly lubricated bronze bearings

= 1.08 for well lubricated bronze bearings

= 1.05 for ball and roller bearings

η – Efficiency of pulley

= 1/C

For a pulley system,

$$P = \frac{C^z(C-1)}{C^z - 1} \times F \text{ for raising the load}$$

where z – Number of times the rope passes over the pulleys

$$P = \frac{(C-1)}{C(C^z - 1)} \times F \text{ for lowering the load}$$

The diameter of the sheaves for fibrous ropes used for lifting purposes should be at least 10d.

4.6 Wire Ropes

Symbols

d – Diameter of the wire rope, mm

d_w – Diameter of the wire, mm

A – Area of the wire rope

D – Diameter of the sheave or drum

P – Load lifed

w – Weight of the rope

σ – Direct stress

$$= \frac{P + w}{A}$$

P_b – Bending load on the rope

$$= \frac{E \times d_w \times A}{D}$$

where E – Modulus of elasticity of the wire rope

$$P_b = \frac{\pi}{4}(d_w)^2 \times n \times \sigma_b$$

where σ_b – Bending stress in each wire

n – Total number of wires in the rope section

P_a – Load on the rope due to acceleration

$$= \frac{P + w}{g} \times a$$

where g – Acceleration due to gravity

a – Acceleration of the load and rope

$$= \frac{v}{60t}$$

t – Time necessary to attain a speed v

P_{st} – Impact load on starting (when there is slackness in the rope)

$$= (P + w)\left[1 + \sqrt{1 + \frac{2a \times h \times E}{\sigma \times l \times g}}\right]$$

where h – Slackness in the rope

l – Length of the rope

P_{st} – Impact load during starting (when there is no slackness in the rope)

$$= 2(P + w)$$

Effective load in the rope during normal working = $P + w + P_b$

Effective load in the rope during starting = $P_{st} + P_b$

Effective load in the rope during acceleration of the load = $P + w + P_b + P_a$

Table 4.12 Steel wire ropes for haulage purposes in mines

Type of rope	Average weight, N/m	Tensile strength, N	
		Tensile strength of wire	
		1600 N/mm²	**1800 N/mm²**
6 × 7	$0.0347\ d^2$ (d in mm)	$530\ d^2$ (d in mm)	$600\ d^2$ (d in mm)
6 × 19	$0.0363\ d^2$ (d in mm)	$530\ d^2$ (d in mm)	$595\ d^2$ (d in mm)

Table 4.13 Steel wire suspension ropes for lifts, elevators and hoists

Type of rope	Average weight, N/m	Tensile strength, N	
		Tensile strength of wire	
		1100-1250 N/mm²	**1250-1400 N/mm²**
6 × 19	$0.0383\ d^2$ (d in mm)	$385\ d^2$ (d in mm)	$435\ d^2$ (d in mm)
8 × 19	$0.034\ d^2$ (d in mm)	$355\ d^2$ (d in mm)	$445\ d^2$ (d in mm)

Table 4.14 Steel wire ropes for general engineering purposes such as cranes, excavators etc.

Type of rope	Average weight, N/m	Average tensile strength, N	
		Tensile strength of wire	
		1600-1750 N/mm²	**1750-1900 N/mm²**
6 × 19	$0.0375\ d^2$ (d in mm)	$540\ d^2$ (d in mm)	$590\ d^2$ (d in mm)
6 × 37	$0.038\ d^2$ (d in mm)	$510\ d^2$ (d in mm)	$550\ d^2$ (d in mm)

Table 4.15 Steel wire ropes used in oil wells

Type of rope	Average weight, N/m	Average tensile strength, N		
		Tensile strength of wire		
		1600-1800 N/mm²	**1800-2000 N/mm²**	**2000-2250 N/mm²**
6 × 7	$0.037\ d^2$ (d in mm)	$550\ d^2$ (d in mm)	$610\ d^2$ (d in mm)	–
6 × 19	$0.037\ d^2$ (d in mm)	$510\ d^2$ (d in mm)	$570\ d^2$ (d in mm)	$630\ d^2$ (d in mm)
6 × 37	$0.037\ d^2$ (d in mm)	$490\ d^2$ (d in mm)	$540\ d^2$ (d in mm)	$600\ d^2$ (d in mm)
8 × 19	$0.0338\ d^2$ (d in mm)	–	$530\ d^2$ (d in mm)	–

Table 4.16 Wire diameter and area of wire rope

Type of wire rope	Wire diameter (d_w)	Area of wire rope (A)
6 × 7	0.106 d (d in mm)	0.38 d^2 (d in mm)
6 × 19	0.063 d (d in mm)	0.38 d^2 (d in mm)
6 × 37	0.045 d (d in mm)	0.38 d^2 (d in mm)
8 × 19	0.050 d (d in mm)	0.35 d^2 (d in mm)

Table 4.17 Factor of safety for wire ropes

Application of wire rope	Factor of safety
Track	4.2
Guys	3.5
Mine hoists :	
Depths upto 150 m	8
300 – 600 m	7
600 – 900 m	6
Over 900 m	5
Miscellaneous hoists	5
Haulage ropes	6
Small electric and air hoists	7
Over head and gantry cranes	6
Jib and pillar cranes	6
Hot ladle cranes	8
Slings	8
Derricks	6

Table 4.18 Sheave diameters for wire ropes

Type of wire rope	Recommended sheave diameter		Uses
	Minimum sheave diameter	Preferred sheave diameter	
6 × 7	42 d	72 d	Mines, haulage tramways
6 × 19	30 d	45 d	Hoisting rope
	60 d	100 d	Cargo cranes, mine hoists
	20 d	30 d	Derricks, dredges, elevators, tramways, well drilling
6 × 37	18 d	27 d	Cranes, high speed elevators
8 × 19	21 d	31 d	Extra flexible hoisting rope

Chapter 5

Spur Gear Drives

Design of Spur Gears based on minimum centre distance and module

Nominal torque transmitted by pinion, $M_t = \dfrac{60 \times 10^6 \times P}{2\pi N_P}$ N-mm

where P – Power transmitted in kW

N_P – Speed of pinion in rpm

Design torque, $[M_t] = M_t \times K\,K_d$ N-mm

K – Load concentration factor

K_d – Dynamic load factor

$K\,K_d = 1.3$

Centre distance based on surface compressive strength of weaker gear in mm,

$$C \geq (V.R \pm 1)\sqrt[3]{\left(\frac{0.74}{[\sigma_c]}\right)^2 \frac{E[M_t]}{V.R \times \psi}} \quad \text{for pressure angle } \phi = 20^\circ$$

$$C \geq (V.R \pm 1)\sqrt[3]{\left(\frac{0.85}{[\sigma_c]}\right)^2 \frac{E[M_t]}{V.R \times \psi}} \quad \text{for pressure angle } \phi = 14\tfrac{1}{2}^\circ$$

(+ for external gears, – for internal gears)

where V.R – Velocity ratio

$= \dfrac{T_G}{T_P} > 1$

T_P – Number of teeth on pinion

T_G – Number of teeth on gear

ψ – Ratio of face width to centre distance (Refer Table 5.1)

$$= \frac{b}{C}$$

b – Face width of the tooth, mm

E – Equivalent young's modulus, N/mm^2

$$= \frac{2E_P E_G}{E_P + E_G}$$

E_P – Young's modulus of pinion material, N/mm^2 (Refer Table 1 in Appendix)

E_G – Young's modulus of gear material, N/mm^2 (Refer Table 1 in Appendix)

$[\sigma_c]$ – Design compressive stress, N/mm^2 (Refer Table 5.2)

If working life specified, $[\sigma_c] = C_B \text{ HB } K_{Cl}$

$$= C_R \text{ HRC } K_{Cl}$$

where C_B or C_R – Coefficient depending on the surface hardness (Refer Table 5.3)

HB or HRC – Brinell or Rockwell 'C' hardness number (Refer Table 5.4)

K_{Cl} – Life factor for compressive strength

$$= \sqrt[6]{\frac{10^7}{N}}$$

where N – Life in number of cycles the pinion tooth has undergone in its life period of H hours

$$N = 60\, N_P H$$

Induced surface compressive stress on gear tooth in N/mm^2,

$$\sigma_C = 0.74\left(\frac{V.R \pm 1}{C}\right)\sqrt{\left(\frac{V.R \pm 1}{V.R \times b}\right)E[M_t]} \le [\sigma_C] \qquad \text{for } \phi = 20^\circ$$

$$\sigma_C = 0.85\left(\frac{V.R \pm 1}{C}\right)\sqrt{\left(\frac{V.R \pm 1}{V.R \times b}\right)E[M_t]} \le [\sigma_C] \qquad \text{for } \phi = 14\tfrac{1}{2}^\circ$$

Module based on beam strength of tooth in mm,

$$m \ge 1.26\sqrt[3]{\frac{[M_t]}{Y[\sigma_b]\psi_m T_P}}$$

where Y – Form factor (Table 5.5)

ψ_m – Ratio of face with to module

$$= \frac{b}{m} \quad \text{(In general } b = 10\ m\text{)}$$

$[\sigma_b]$ – Design bending stress for the weaker material, N/mm^2

$$= \frac{1.4k_{bl}}{nk_\sigma} \times \sigma_{-1}$$

k_{bl} – Life factor bending (Table 5.6)

σ_{-1} – Endurance limit stress in bending for complete reversal of stresses, N/mm^2 (Table 5.7)

n – Factor of safety (Table 5.8)

k_σ – Stress concentration factor for the fillet (Table 5.9)

For standard modules, refer Table 5.10.

$$T_P = \frac{2C}{(V.R + 1)m}$$

Bending stress, $\sigma_b = \frac{V.R \pm 1}{CmbY}[M_t] \le [\sigma_b]$

Check for plastic deformation

$$\sigma_{c\,max} = \sigma_c \sqrt{\frac{(M_t)_{max}}{M_t}} \le [\sigma_c]_{max}$$

$$\sigma_{b\,max} = \sigma_b \frac{(M_t)_{max}}{M_t} \le [\sigma_b]_{max}$$

M_t – Nominal twisting moment on pinion

$(M_t)_{max}$ – Instantaneous maximum twisting moment (starting, braking, sudden stopping etc.,)

$= 2M_t$ can be assumed

$\sigma_{c\,max}$ – Maximum surface compressive stress due to $(M_t)_{max}$, N/mm^2

$[\sigma_c]_{max}$ – Maximum allowable surface contact stress to avoid plastic deformation or brittle crushing (Refer Table 5.11)

$\sigma_{b_{max}}$ – Maximum bending stress due to $(M_t)_{max}$, N/mm^2

$[\sigma_b]_{max}$ – Maximum allowable bending stress to avoid plastic deformation or brittle crushing (Refer Table 5.11)

Design of Spur Gears as recommended by AGMA procedure

Design tangential tooth load,

$$F_t = \sigma_w b \pi m y$$

where m – Module in mm and b – Face width of the tooth in mm

$$F_t = \frac{P}{v} \times C_s$$

P – Power transmitted in watts

v – Pitch line velocity in m/s

$$= \frac{\pi D N}{60 \times 1000} \text{ m/s}$$

D – Pitch circle diameter, mm

N – Speed, rpm

C_s – Service factor (Refer Table 5.12)

σ_w – Permissible working stress for gear tooth, N/mm^2

$$= \sigma_o \times C_v$$

σ_o – Allowable static stress (Refer Table 5.13)

C_v – Velocity factor

$= \frac{3}{3+v}$ for ordinary cut gears operating at velocities upto 12.5 m/s

$= \frac{4.5}{4.5+v}$ for carefully cut gears operating at velocities upto 12.5 m/s

$= \frac{6}{6+v}$ for very accurately cut and ground metallic gears

$= \frac{0.75}{0.75+\sqrt{v}}$ for precision gears upto 20 m/s

$= \left(\frac{0.75}{1+v}\right) + 0.25$ for non-metallic gears y – Lewis form factor

$= 0.124 - \frac{0.684}{T}$ for $14\frac{1}{2}^{\circ}$ composite and full depth involute systems

$= 0.154 - \frac{0.912}{T}$ for 20° full depth involute system

$= 0.175 - \frac{0.841}{T}$ for 20° stub involute system

T – Number of teeth

Static tooth load (Endurance strength of the tooth),

$$F_s = \sigma_e b \pi m y$$

where σ_e – Flexural endurance limit, N/mm^2 (Refer Table 5.14)

Dynamic tooth load, $F_d = F_t + \frac{21v(bC' + F_t)}{21v + \sqrt{bC' + F_t}}$

where steady transmitted load, $F_t = \frac{P}{v}$

A deformation or dynamic factor, $C' = \frac{K'e}{\frac{1}{E_P} + \frac{1}{E_G}}$ (Refer Table 5.15)

where $K' = 0.107$ for $14\frac{1}{2}^{\circ}$ full depth involute system

$= 0.111$ for 20° full depth involute system

$= 0.115$ for 20° stub system

e – Tooth error in action, mm (Refer Table 5.16)

For safety against breakage, $F_s > F_d$

Wear tooth load,

$$F_w = D_P b Q K$$

where D_P – Pitch circle diameter of the pinion, mm

Q – Ratio factor

$$= \frac{2 \times V.R}{V.R \pm 1}$$

$$Q = \frac{2T_G}{T_G + T_P} \text{ for external gears}$$

$$= \frac{2T_G}{T_G - T_P} \text{ for internal gears}$$

K – Load-stress factor, N/mm^2

$$= \frac{(\sigma_{es})^2 \sin\phi}{1.4}\left(\frac{1}{E_P} + \frac{1}{E_G}\right)$$

where σ_{es} – Surface endurance limit, N/mm^2 (Refer Table 5.14)

For satisfactory wear of gear teeth, $F_w > F_d$

Velocity ratio, $V.R = \dfrac{T_G}{T_P}$ $\left(\dfrac{T_G}{T_P} = \dfrac{D_G}{D_P} = \dfrac{N_P}{N_G}\right)$

where D_G – Pitch circle diameter of the gear, mm

N_G – Speed of gear, rpm

Pitch circle diameter for pinion, $D_P = m\,T_P$

Pitch circle diameter for gear, $D_G = m\,T_G$

Face width of gears, b = 10 m

For standard proportions of spur gears, refer Table 5.17.

Minimum number of teeth on the pinion in order to avoid interference:

12 (for $14\frac{1}{2}^\circ$ composite system)

32 (for $14\frac{1}{2}^\circ$ full depth involute system)

18 (for 20° full depth involute system)

14 (for 20° stub involute system)

Table 5.1 Ratio of face width to centre distance

Type of gear transmission	$\psi = \frac{b}{C}$
Open type gearing	0.1 to 0.3
Speed reducers (Closed type)	
(a) High speed 8 to 25 m/s	upto 0.3
(b) Medium speed 3 to 8 m/s	upto 0.6
(c) Low speed 1 to 3 m/s	upto 1.0
Gear boxes with sliding gears	0.12 to 0.15

Recommended ψ values, ψ = 0.20, 0.25, 0.30, 0.40, 0.50, 0.60, 0.80, 1.0, 1.2

Table 5.2 Design compressive stress for gear materials

IS Classification		σ_u N/mm^2	Endurance limit σ_e min. N/mm^2	Number of cycles	Surfaces hardness HB	$[\sigma_b]$ N/mm^2	$[\sigma_c]$ N/mm^2
Cast iron	Grade 20	≥ 200	100	10^7	179-223	46-50	500
	Grade 25	≥ 250	120	10^7	197-241	55-60	600
	Grade 35	≥ 350	130	10^7	207-241	55-60	600
	Grade 35 Heat treated	≥ 350	160	10^7	300 min.	75-80	750
Steel	C 45	≥ (630)	270	10^7	175-215	135-140	500
	15 Ni 2 Cr 1 Mo 15 Carburized	≥ 900	550	25×10^7	case (500) core (250)	300-320	950
	40 Ni 2 Cr 1 Mo 28 Case hardened	≥ 1550	600	25×10^7	case (600) core (250)	380-400	1100

σ_u, Ultimate tensile strength

σ_e min., Minimum endurance limit stress for complete reversal of stresses

$[\sigma_c]$, Design surface compressive stress

Table 5.3 Coefficients C_B and C_R

Wheel material	Heat treatment	Surface hardness	Coefficient C_B or C_R
Carbon steels and alloy steel or any type	Normalised or hardened and tempered	HB ≤ 350	$C_B = 2.5$
High strength alloy nickel chromium steels	Case hardened	HRC = 55 to 63	$C_R = 31$
Alloy steels	Case hardened	HRC = 55 to 63	$C_R = 28$
Carbon & manganese steels C15; C20; C15 Mn85; C20 Mn85	Case hardened	HRC = 55 to 63	$C_R = 22$
Alloy steels, carbon steels C40; C45	Hardened and tempered	HRC = 40 to 55	$C_R = 26.5$
Alloy steels, carbon steels C40; C45	Surface hardened	HRC = 40 to 55	$C_R = 23$
Cast iron, grade 20, 25	–	HB = 170 to 200	$C_B = 2$
Cast iron, grade 30, 35	–	HB = 200 to 260	$C_B = 2.3$

Table 5.4 Properties of gear materials

Material	IS Specification	Tensile strength, N/mm^2	Brinell hardness number
Grey Cast Iron	Grade 25	250	197
Phosphor Bronze	Sand Cast Chill Cast Centrifugally Cast	160 240 270	60 70 90
Cast steel	Grade 1	550	145
Constructional Steel	St 58	580-680	160-190
Plain Carbon Steel	C 60	600-700	180-200
Carbon steel for surface hardening	C 45 C 55	700 720	145 core 460 case 200 core 520 case
Carbon steel for case hardening	C 14	500-750	650 case
Direct hardening Alloy Steels	40 Ni 2 Cr 1 Mo 28 30 Ni 4 Cr I 35 Ni 1 Cr 60 40 Ni 3	1550 1000-1150 1150 750-1050	444 400-500 400-500 400-500
Alloy Steels for case hardening	17 Mn 1 Cr 95 15 Ni 2 Cr 1 Mo 15 13 Ni 3 Cr 80 15 Ni 4 Cr 1 15 Cr 65	800-1100 1040 900-1200 1200-1500 600-850	650 case 630 case 600-620 case 600-650 case 650 case
Nitriding Alloy steel	40 Cr 2 Al 1 Mo 18	> 660	750-800 case

Table 5.5 Form factor Y

Number of Teeth	$\phi = 14\frac{1}{2}^\circ$	$\phi = 20^\circ$	$\phi = 20^\circ$ stub
10	0.176	0.201	0.261
11	0.192	0.226	0.289
12	0.210	0.245	0.311
13	0.233	0.264	0.324
14	0.236	0.276	0.339
15	0.245	0.289	0.349
16	0.255	0.295	0.360
17	0.264	0.302	0.368
18	0.270	0.308	0.377
19	0.277	0.314	0.386
20	0.283	0.320	0.393
21	0.289	0.326	0.399
22	0.292	0.330	0.404
23	0.296	0.333	0.408
24	0.302	0.337	0.411
25	0.305	0.340	0.416
26	0.308	0.344	0.421
27	0.311	0.348	0.426
28	0.314	0.352	0.430
29	0.316	0.355	0.434
30	0.318	0.358	0.437
32	0.322	0.364	0.443
33	0.324	0.367	0.445
35	0.327	0.373	0.449
37	0.330	0.380	0.454
39	0.335	0.386	0.457
40	0.336	0.389	0.459
45	0.340	0.399	0.468
50	0.346	0.408	0.474
55	0.352	0.415	0.480
60	0.355	0.421	0.484
65	0.358	0.425	0.488
70	0.360	0.429	0.493
75	0.361	0.433	0.496
80	0.363	0.436	0.499
90	0.366	0.442	0.503
100	0.368	0.446	0.506
150	0.375	0.458	0.518
200	0.378	0.463	0.524
300	0.382	0.471	0.534

Table 5.6 Life factor for bending, k_{bl}

Material	Brinell Hardness	Life (cycles)	k_{bl}
Steel	≤ 350	$\geq 10^7$	1.0
Steel	≤ 350	$< 10^7$	$\{10^7/N\}^{1/9}$
Steel	> 350	$\geq 25 \times 10^7$	0.70
Steel	> 350	$< 25 \times 10^7$	$\{10^7/N\}^{1/9}$
Cast iron	–	–	$\{10^7/N\}^{1/9}$

Table 5.7 Endurance limit stress in bending, σ_{-1}

Material	Endurance limit stress in bending, N/mm^2
Forged steels	$0.25\,(\sigma_u + \sigma_y) + 50$
Cast steels	$0.22\,(\sigma_u + \sigma_y) + 50$
Alloy steels	$0.35\,\sigma_u + 120$
Cast iron	$0.45\,\sigma_u$

Table 5.8 Factor of Safety, n

Material	Mode of manufacture	Heat treatment	n
Steel, cast iron	Cast	Not heat treatment	2.5
Steel, cast iron	Cast	Tempered or normalized	2.0
Steel	Cast or forged	Case hardened	2.0
Steel	Forged	Surface hardened	2.5
Steel	Forged	Normalized	2.0

Table 5.9 Stress concentration factor for the fillet, k_σ

Material and treatment	k_σ
Steel-normalized surface hardened	1.5
Steel-case hardened	1.2
Cast iron	1.2

Table 5.10 Standard modules in mm

Modules in mm	
Choice 1 (Preferred)	1.0, 1.25, 1.5, 2.0, 2.5, 3.0, 4.0, 5.0, 6.0, 8.0, 10.0, 12.0, 16.0, 20.0
Choice 2	1.125, 1.375, 1.75, 2.25, 2.75, 3.5, 4.5, 5.5, 7.0, 9.0, 11.0, 14.0, 18.0
Choice 3	3.25, 3.75, 6.5

Table 5.11 Maximum allowable surface contact and bending stresses

Material	Surface hardness	$[\sigma_c]_{max}$, N/mm^2	$[\sigma_b]_{max}$, N/mm^2
Steel	HB $\leq$ 350	3.1 σ_y	0.8 σ_y
	HB > 350	42 HRC	$0.36 \frac{\sigma_u}{k_\sigma}$
Cast Iron	HB $\leq$ 350	1.8 σ_u	0.6 σ_u

σ_y – yield stress in tension, N/mm^2

σ_u – ultimate tensile strength, N/mm^2

Table 5.12 Values of service factor, C_S

Type of load	Type of service		
	Intermittent or 3 hours per day	**8-10 hours per day**	**Continuous 24 hours per day**
Steady	0.8	1.00	1.25
Light shock	1.00	1.25	1.54
Medium shock	1.25	1.54	1.80
Heavy shock	1.54	1.80	2.00

Table 5.13 Values of allowable static stress, σ_o

Material	Allowable static stress N/mm^2
Cast iron, ordinary	56
Cast iron, medium grade	70
Cast iron, highest grade	105
Cast steel, untreated	140
Cast steel, heat treated	196
Forged carbon steel-case hardened	126
Forged carbon steel-untreated	140 to 210
Forged carbon steel-heat treated	210 to 245
Alloy steel-case hardened	350
Alloy steel-heat treated	455 to 472
Phosphor bronze	84
Non-metallic materials	
Rawhide, fabroil	42
Bakelite, Micarta, celoron	56

Table 5.14 Values of flexural and surface endurance limits

Material of pinion and gear	Brinell hardness number	Flexural endurance limit σ_e, N/mm^2	Surface endurance limit σ_{es}, N/mm^2
Grey cast iron	160	84	630
Semi-steel	200	126	630
Phosphor bronze	100	168	630
Steel	150	252	350
	200	350	490
	240	420	616
	280	490	721
	300	525	770
	320	560	826
	350	595	910
	360	630	–
	400 and above	700	1050

Table 5.15 Values of deformation factor

Material		Involute tooth form	Values of deformation factor C' in N/mm				
			Tooth error in action (e) in mm				
Pinion	Gear		0.01	0.02	0.04	0.06	0.08
Cast iron	Cast iron	$14\frac{1}{2}^\circ$	55	110	220	330	440
Steel	Cast iron		76	152	304	456	608
Steel	Steel		110	220	440	660	880
Cast iron	Cast iron	20° full depth	57	114	228	342	456
Steel	Cast iron		79	158	316	474	632
Steel	Steel		114	228	456	684	912
Cast iron	Cast iron	20° stub	59	118	236	354	472
Steel	Cast iron		81	162	324	486	648
Steel	Steel		119	238	476	714	952

Table 5.16 Values of tooth error in action

Module (m) in mm	Tooth error in action (e) in mm		
	First class commercial gears	**Carefully cut gears**	**Precision gears**
Upto 4	0.051	0.025	0.0125
5	0.055	0.028	0.015
6	0.065	0.032	0.017
7	0.071	0.035	0.0186
8	0.078	0.0386	0.0198
9	0.085	0.042	0.021
10	0.089	0.0445	0.023
12	0.097	0.0487	0.0243
14	0.104	0.052	0.028
16	0.110	0.055	0.030
18	0.114	0.058	0.032
20	0.117	0.059	0.033

Table 5.17 Standard proportions of spur gears

Particulars	$14\frac{1}{2}^{\circ}$ composite or full depth involute systems	20° full depth involute system	20° stub involute system
Addendum	1 m	1 m	0.8 m
Dedendum	1.25 m	1.25 m	1 m
Working depth	2 m	2 m	1.60 m
Minimum total depth	2.25 m	2.25 m	1.80 m
Tooth thickness	1.5708 m	1.5708 m	1.5708 m
Minimum clearance	0.25 m	0.25 m	0.2 m
Fillet radius at root	0.4 m	0.4 m	0.4 m

Chapter 6

Helical and Bevel Gear Drives

6.1 Helical Gears

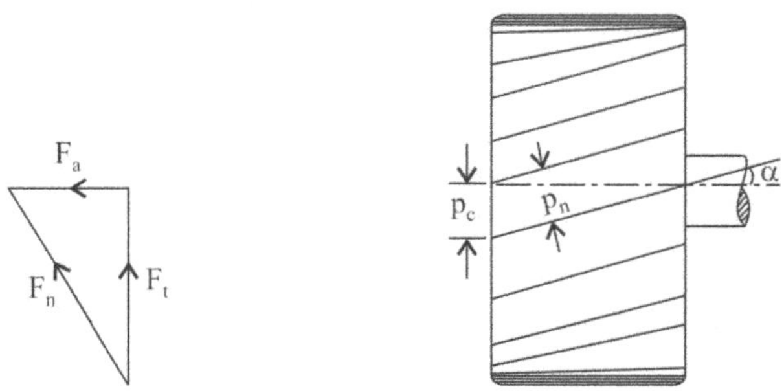

Fig. 6.1 Helical gear

Normal pitch, $p_n = p_c \cos \alpha$

where

p_c – Axial pitch.

α – Helix angle

The relation between normal pressure angle (ϕ_N) in the normal plane and pressure angle (ϕ) in the diametral plane (plane of rotation) is

$\tan\phi_N = \tan \phi \cos \alpha$

Also $F_a = F_n \sin \alpha = F_t \tan \alpha$

F_n – Normal tooth load

F_a – Axial component

F_t – Tangential component

Equivalent number of tooth for helical gears, $T_E = \dfrac{T}{\cos^3 \alpha}$

T – Actual number of teeth on helical gear

Strength of Helical Gears

Tangential tooth load, $F_t = (\sigma_o \times C_v)\, b\, \pi\, m_n\, y'$

where

m_n – Normal module, mm

$= m \cos \alpha$

m – Module, mm

b – Face width, mm

σ_o – Allowable static stress (Refer Table 5.13)

C_v – Velocity factor

$= \dfrac{6}{6 + v}$ for peripheral velocities from 5 m/s to 10 m/s

$= \dfrac{15}{15 + v}$ for peripheral velocities from 10 m/s to 20 m/s

$= \dfrac{0.75}{0.75 + \sqrt{v}}$ for peripheral velocities greater than 20 m/s

$= \dfrac{0.75}{1 + v} + 0.25$ for non-metallic gears

v – Peripheral velocity, m/s

y' – Tooth form factor or Lewis factor corresponding to the equivalent number of teeth

$y' = 0.124 - \dfrac{0.684}{T_E}$ for $14\frac{1}{2}^{\circ}$ teeth

$= 0.154 - \dfrac{0.912}{T_E}$ for 20° full depth involute system

$= 0.175 - \dfrac{0.841}{T_E}$ for 20° stub system

Static tooth load, $F_s = \sigma_e\, b\, \pi\, m_n\, y'$

where σ_e – Flexural endurance limit, N/mm^2 (Refer Table 5.14)

Dynamic tooth load, $F_d = F_t + \dfrac{21v\left(bC'\cos^2\alpha + F_t\right)\cos\alpha}{21v + \sqrt{bC'\cos^2\alpha + F_t}}$

where F_t, v and C' have same meanings as in spur gears.

Maximum wear tooth load, $F_w = \dfrac{D_P\, bQK}{\cos^2\alpha}$

where D_P, b and Q have same meanings as discussed in spur gears.

$$K = \frac{(\sigma_{es})^2 \sin\phi_N}{1.4}\left[\frac{1}{E_P} + \frac{1}{E_G}\right]$$

σ_{es} – Surface endurance limit, N/mm^2 (Refer Table 5.14)

E_P, E_G are same as in spur gears.

$$F_s > F_d$$

$$F_w > F_d$$

Proportions for Helical Gears Recommended by American Gear Manufacturers' Association (AGMA)

Helix angle, $\alpha = 20^\circ$ to 45°

Addendum = 0.8 m (Maximum)

Dedendum = 1 m (Minimum)

Minimum total depth = 1.8 m

Minimum clearance = 0.2 m

Thickness of tooth = 1.5708 m

6.2 Bevel Gears

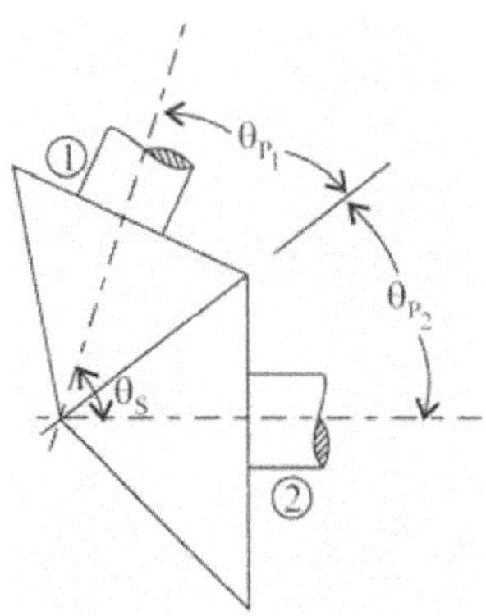

Fig. 6.2 Bevel gears

Miter gears: Equal teeth and equal pitch angles, shaft axes intersect at right angle.

θ_{P_1} – Pitch angle of Pinion

θ_{P_2} – Pitch angle of gear

θ_S – Angle between the shaft axes

D_P – Pitch diameter of pinion

D_G – Pitch diameter of gear

$$V.R = \frac{D_G}{D_P} = \frac{T_G}{T_P} = \frac{N_P}{N_G}$$

When $\theta_S = 90^\circ$,

$$\theta_{P_1} = \tan^{-1}\left(\frac{D_P}{D_G}\right) = \tan^{-1}\frac{1}{V.R}$$

$$\theta_{P_2} = \tan^{-1}(V.R)$$

θ_P – Pitch angle of cone

R – Pitch circle radius of Bevel pinion or gear

R_B – Back cone distance or equivalent pitch circle radius of spur pinion or gear

$$R_B = R \sec \theta_P$$

Equivalent number of teeth, $T_E = \frac{2R_B}{m} = \frac{2R \sec \theta_P}{m} = T \sec \theta_P$

where

m – Module

T – Actual number of teeth on the gear

Strength of Bevel Gears

$$F_t = (\sigma_o \times C_v)\, b \pi\, m y' \left(\frac{L-b}{L}\right)$$

$$C_v = \frac{3}{3+v}, \text{ cut by form cutters}$$

$$= \frac{6}{6+v}, \text{ cut by precision cutters}$$

y' – Tooth factor for equivalent teeth

$$= 0.124 - \frac{0.684}{T_E} \text{ for } 14\tfrac{1}{2}^\circ \text{ teeth}$$

$$= 0.154 - \frac{0.912}{T_E} \text{ for } 20^\circ \text{ full depth involute system}$$

$$= 0.175 - \frac{0.841}{T_E} \text{ for } 20^\circ \text{ stub system}$$

L – Slant height of pitch cone

$$= \sqrt{\left(\frac{D_G}{2}\right)^2 + \left(\frac{D_P}{2}\right)^2}$$

Generally L/b is 3 and b is 6.3 m to 9.5 m.

$$\text{Minimum number of teeth on pinion} = \frac{48}{\sqrt{1+(V.R)^2}}$$

where V.R is the required velocity ratio.

$$\text{Dynamic load,}\quad F_d = F_t + \frac{21v(bC'+F_t)}{21v+\sqrt{bC'+F_t}}$$

where F_t, v and C' have same meanings as in spur gears.

$$\text{Static tooth load,}\quad F_s = \sigma_e\, b\, \pi\, m\, y' \left(\frac{L-b}{L}\right)$$ (For σ_e values, refer Table 5.14)

$$\text{Maximum limiting load for wear,}\quad F_w = \frac{D_P b Q K}{\cos\theta_{P_1}}$$

D_P, b, K have same meaning as that of spur gears.

$$Q = \frac{2T_{EG}}{T_{EG} + T_{EP}}$$

where

T_{EG}, T_{EP} are equivalent number of teeth for gear and pinion respectively.

$$F_s > F_d$$

$$F_w > F_d$$

Forces

$F_t = F_n \cos \phi$

$F_r = F_n \sin \phi = F_t \tan \phi$

where

ϕ – Pressure angle

F_n – Normal force on the tooth

F_t – Tangential component

F_r – Radial component

Proportions for Bevel Gears

Addendum = 1 m

Dedendum = 1.2 m

Clearance = 0.2 m

Working depth = 2 m

Thickness of tooth = 1.5708 m

Chapter 7

Power Screws

Symbols

α – Lead angle

ϕ – Friction angle

W – Load to be lifted

L' – Lead

d – Pitch diameter of screw

μ – Coefficient friction for screw rod

d_c – Root diameter of screw rod

d_o – Major diameter of threads for the nut

L – Equivalent length of the screw rod

μ_c – Coefficient of friction of collar

R_1 – Outside diameter of collar

R_2 – Inside diameter of collar

k – Least radius of gyration

n – Number of threads in contact with the nut $= \dfrac{\text{Height of the nut}}{\text{Pitch of threads}}$

t – Thickness of thread

Design equations

From wear consideration,

Pitch circle diameter of nut, $d \geq \sqrt{\dfrac{2W}{\pi\psi[p_b]}}$

where $[p_b]$ – Allowable bearing pressure (Refer Table 7.1)

ψ – Nut height factor $= \frac{H}{d}$

= 1.2 to 2.5 for solid nuts

= 2.5 to 3.5 for split nuts

H – Height of nut (i.e., nut thickness)

For standard dimensions of screw and nut, refer Tables 7.2, 7.3, 7.4 & 7.5.

Torque required to raise the load	$T_r = \frac{Wd}{2}\tan(\alpha + \phi)$
Torque required to lower the load	$T_l = \frac{Wd}{2}\tan(\phi - \alpha)$
	$\alpha = \tan^{-1}\frac{L'}{\pi d}$
	$\phi = \tan^{-1}\mu$
Efficiency of the power screw	$\eta = \frac{\tan\alpha}{\tan(\alpha + \phi)}$
Axial stress for short screw rod	$\sigma_a = \frac{4W}{\pi d_c^2}$
Axial stress for long screw rod	$\sigma_a = \frac{4W\left[1 + a\left(\frac{L}{k}\right)^2\right]}{\pi d_c^2}$
	$a = \frac{1}{7500}$ for steel
Torsional shear stress in the screw rod	$\tau = \frac{16T_r}{\pi d_c^3}$
Maximum principal stress in the screw-rod	$\sigma = \frac{1}{2}\left[\sigma_a + \sqrt{\sigma_a^2 + 4\tau^2}\right]$
Maximum shear stress	$\tau_m = \frac{1}{2}\sqrt{\sigma_a^2 + 4\tau^2}$
Direct shear stress induced in the threads	$\tau = \frac{W}{n\pi d_c t}$ or $\frac{W}{n\pi d_o t}$ (for nut)
Bearing pressure	$p_b = \frac{W}{n\pi dt}$
Torque required to overcome collar friction (For uniform pressure)	$T = \frac{2}{3}\mu_c W\left[\frac{R_1^3 - R_2^3}{R_1^2 - R_2^2}\right]$
Torque required to overcome collar friction (For uniform wear)	$T = \mu_c W\left(\frac{R_1 + R_2}{2}\right)$

Table 7.1 Values of allowable bearing pressure

Material combination	Allowable bearing pressure $[p_b]$, N/mm^2
Steel screw-Bronze nut	12
Steel screw-Cast iron nut	8
Steel screw-Bronze nut for machine tool lead screws	3

Table 7.2 Basic dimensions for square threads in mm (fine series) according to IS: 4694

Nominal diameter (d_1)	Major diameter		Minor diameter (d_c)	Pitch (p)	Depth of thread		Area of core (A_c) mm^2
	Bolt (d)	Nut (D)			Bolt (h)	Nut (H)	
10	10	10.5	8	2	1	1.25	50.3
12	12	12.5	10				78.5
14	14	14.5	12				113
16	16	16.5	14				154
18	18	18.5	16				201
20	20	20.5	18				254
22	22	22.5	19	3	1.5	1.75	284
24	24	24.5	21				346
26	26	26.5	23				415
28	28	28.5	25				491
30	30	30.5	27				573
32	32	32.5	29				661
(34)	34	34.5	31				755
36	36	36.5	33				855
(38)	38	38.5	35				962
40	40	40.5	37				1075
42	42	42.5	39				1195
44	44	44.5	41				1320
(46)	46	46.5	43				1452
48	48	48.5	45				1590
50	50	50.5	47				1735
52	52	52.5	49				1886
55	55	55.5	52	3	1.5	1.75	2124
(58)	58	58.5	55				2376
60	60	60.5	57				2552
(62)	62	62.5	59				2734
65	65	65.5	61	4	2	2.25	2922
(68)	68	68.5	64				3217
70	70	70.5	66				3421
(72)	72	72.5	68				3632
75	75	75.5	71				3959
(78)	78	78.5	74				4301
80	80	80.5	76				4536

Table 7.2 ***Contd...***

(82)	82	82.5	78				4778
(85)	85	85.5	81				5153
(88)	88	88.5	84				5542
90	90	90.5	86				5809
(92)	92	92.5	88	4	2	2.25	6082
95	95	95.5	91				6504
(98)	98	98.5	94				6960
100	100	100.5	96				7238
(105)	105	105.5	101				8012
110	110	110.5	106				8825
(115)	115	115.5	109				9331
120	120	120.5	114				10207
(125)	125	125.5	119				11122
130	130	130.5	124				12076
(135)	135	135.5	129				13070
140	140	140.5	134				14103
(145)	145	145.5	139	6	3	3.25	15175
150	150	150.5	144				16286
(155)	155	155.5	149				17437
160	160	160.5	154				18627
(165)	165	165.5	159				19856
170	170	170.5	164				21124
(175)	175	175.5	169				22432

Note: Diameters within brackets are of second preference.

Table 7.3 Basic dimensions for square threads in mm (normal series) according to IS: 4694

Nominal diameter (d_1)	Major diameter		Minor diameter (d_c)	Pitch (p)	Depth of thread		Area of core (A_c) mm^2
	Bolt (d)	Nut (D)			Bolt (h)	Nut (H)	
22	22	22.5	17				227
24	24	24.5	19				284
26	26	26.5	21	5	2.5	2.75	346
28	28	28.5	23				415
30	30	30.5	24				452
32	32	32.5	26	6	3	3.25	531
(34)	34	34.5	28				616
36	36	36.5	30				707
(38)	38	38.5	31				755
40	40	40.5	33	7	3.5	3.75	855
(42)	42	42.5	35				962
44	44	44.5	37				1075
(46)	46	46.5	38				1134
48	48	48.5	40	8	4	4.25	1257
50	50	50.5	42				1385
52	52	52.5	44				1521

Table 7.3 ***Contd...***

55	55	55.5	46				1662
(58)	58	58.5	49	9	4.5	4.75	1886
(60)	60	60.5	51				2043
(62)	62	62.5	53				2206
65	65	65.5	55				2376
(68)	68	68.5	58	10	5	5.25	2642
70	70	70.5	60				2827
(72)	72	72.5	62				3019
75	75	75.5	65				3318
(78)	78	78.5	68				3632
80	80	80.5	70				3848
82)	82	82.5	72				4072
85	85	85.5	73	12	6	6.25	4185
(88)	88	88.5	76				4536
90	90	85.5	78				4778
(92)	92	92.5	80				5027
95	95	95.5	83				5411
(98)	98	98.5	86	12	6	7.25	5809
100	100	100.5	88				6082
(105)	105	105.5	93				6793
110	110	110.5	98				7543
(115)	115	116	101				8012
120	120	121	106				882
(125)	125	126	111	14	7	7.5	9677
130	130	131	116				10568
(135)	135	136	121				11499
140	140	141	126				12469
(145)	145	146	131				13478
150	150	151	134				14103
(155)	155	156	139				15175
160	160	161	144				16286
(165)	165	166	149	16	8	8.5	17437
170	170	171	154				18627
(175)	175	176	159				19856

Note: Diameters within brackets are of second preference.

Table 7.4 Basic dimension for square threads in mm (coarse series) according to IS: 4694

Nominal diameter (d_1)	Major diameter		Minor diameter (d_c)	Pitch (p)	Depth of thread		Area of core (A_c) mm^2
	Bolt (d)	Nut (D)			Bolt (h)	Nut (H)	
22	22	22.5	14	8	4	4.25	164
24	24	24.5	16				204
26	26	26.5	18				254
28	28	28.5	20				314
30	30	30.5	20	10	5	5.25	314
32	32	32.5	22				380
(34)	34	34.5	24				452
36	36	36.5	26				531
(38)	38	38.5	28				616
40	40	40.5	28	12	6	6.25	616
(42)	42	42.5	30				707
44	44	44.5	32				804
(46)	46	46.5	34				908
48	48	48.5	36				1018
50	50	50.5	38				1134
52	52	52.5	40				1257
55	55	56	41	14	7	7.25	1320
(58)	58	59	44				1521
60	60	61	46				1662
(62)	62	63	48				1810
65	65	66	49	16	8	8.5	1886
(68)	68	69	52				2124
70	70	71	54				2290
(72)	72	73	56				2463
75	75	76	59				2734
(78)	78	79	62				3019
80	80	81	64				3217
(82)	82	83	66				3421
85	85	86	67	18	9	9.5	3526
(88)	88	89	70				3848
90	90	91	72				4072
(92)	92	93	74				4301
95	95	96	77				4657
(96)	96	99	80				5027

Table 7.4 ***Contd...***

100	100	101	80				5027
(105)	105	106	85	20	10	10.5	5675
110	110	111	90				6362
(115)	115	116	93				6793
120	120	121	98				7543
(125)	125	126	103	22	11	11.5	8332
130	130	131	108				9161
(135)	135	136	111				9667
140	140	141	116	24	12	12.5	10568
(145)	145	146	121				11499
150	150	151	126				12469
(155)	155	156	131				13478
160	160	161	132				13635
(165)	165	166	137				14741
170	170	171	142	28	14	14.5	15837
(175)	175	176	147				16972

Note: Diameters within brackets are of second preference.

Table 7.5 Basic dimensions for trapezoidal/acme threads

Nominal or major diameter (d) mm	Minor or core diameter (d_c) mm	Pitch (p) mm	Area of core (A_c) mm^2
10	6.5	3	33
12	8.5		57
14	9.5		71
16	11.5	4	105
18	13.5		143
20	15.5		189
22	16.5		214
24	18.5	5	269
26	20.5		330
28	22.5		389
30	23.5		434
32	25.5	6	511
34	27.5		594
36	29.5		683
38	30.5		731
40	32.5	7	830
42	34.5		935
44	36.5		1046

Table 7.5 ***Contd...***

46	37.5	8	1104
48	39.5		1225
50	41.5		1353
52	43.5		1486
55	45.5	9	1626
58	48.5		1847
60	50.5		2003
62	52.5	10	2165
65	54.5		2333
68	57.5		2597
70	59.5		2781
72	61.5		2971
75	64.5		3267
78	67.5		3578
80	69.5		3794
82	71.5		4015
85	72.5	12	4128
88	75.5		4477
90	77.5		4717
92	79.5		4964
95	82.5		5346
98	85.5		5741
100	87.5		6013
105	92.5		6720
110	97.5		7466
115	100	14	7854
120	105		8659
125	110		9503
130	115		10387
135	120		11310
140	125		12272
145	130		13273
150	133	16	13893
155	138		14957
160	143		16061
165	148		17203
170	153		18385
175	158		19607

Chapter 8

Worm Gears

Terms used in Worm Gears

1. **Axial pitch:** Pitch measured parallel to axis of the worm, $p_a = p_c$

   p_c – Circular pitch of gear

2. **Lead:** $l = p_a \times n$ for multiple start threads (n is the number of starts)

3. **Lead angle (λ):** Angle between tangent to helix on the pitch cylinder and the plane normal to the axis of the worm

$$\tan\lambda = \frac{l}{\pi D_W} = \frac{p_a\, n}{\pi D_W} = \frac{p_c\, n}{\pi D_W}$$

$$= \frac{\pi\, m\, n}{\pi\, D_W} = \frac{m\, n}{D_W}$$

   where

   m – Module

   D_W – Pitch circle diameter of worm

   For compact design, $\tan\lambda = \left(\frac{N_G}{N_W}\right)^{1/3}$

   where

   N_G – Speed of worm gear

   N_W – Speed of worm

4. **Normal pitch:** $p_N = p_a \cos\lambda$

   Normal lead, $l_N = l \cos\lambda$

5. **Helix angle (α_W):** Angle between tangent to helix and axis of the worm

$$\alpha_W + \lambda = 90°$$

6. Velocity Ratio:

$$V.R = \frac{N_W}{N_G} = \frac{T_G}{n}$$

where

T_G – Number of teeth on the worm gear

7. Efficiency:

$$\eta = \frac{\tan\lambda\,(\cos\phi - \mu\tan\lambda)}{\cos\phi\tan\lambda + \mu}$$

where

ϕ – Normal pressure angle

μ – Coefficient of friction

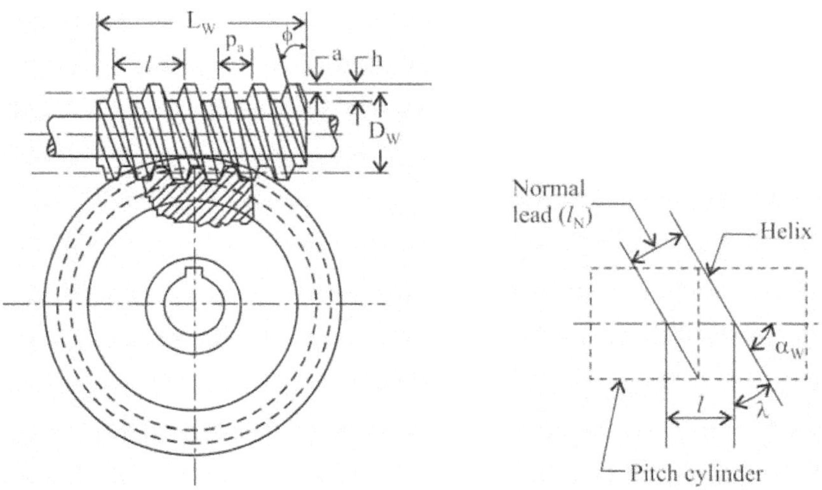

Fig. 8.1 Worm and worm gear

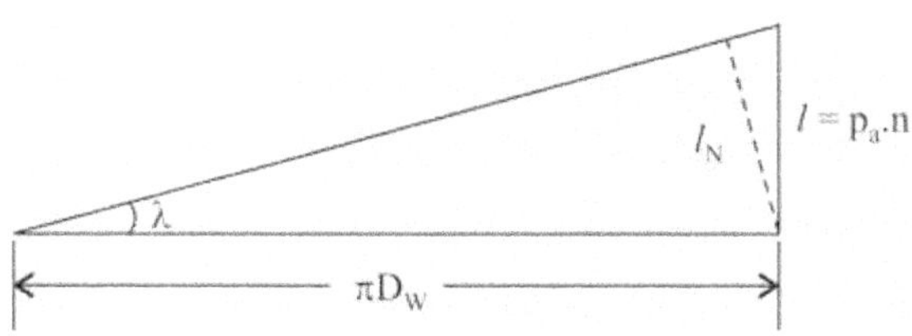

Fig. 8.2 Development of a helix thread

Strength of worm gear teeth

Lewis equation $F_t = (\sigma_o\ C_v)\ b\ \pi\ m\ y$ $C_v = \frac{6}{6+v}$	v – Peripheral velocity of the worm gear in m/s	σ_o – Allowable static stress (Refer Table 5.13)
	$y = 0.124 - \frac{0.684}{T_G}$ for $14\frac{1}{2}^o$ involute teeth $= 0.154 - \frac{0.912}{T_G}$ for 20° involute teeth	
Dynamic tooth load	$F_d = \frac{F_t}{C_v} = F_t\left(\frac{6+v}{6}\right)$	(Not very severe due to sliding action)
Static tooth load	$F_S = \sigma_e\ b\ \pi\ m\, y$	σ_e – Flexural endurance limit, N/mm^2 (Refer Table 5.14)
Wear tooth load	$F_W = D_G\ b\ K$ D_G – Pitch circle diameter of the worm gear	b – Face width of the worm gear K – Load stress factor (Refer table 8.1)
Thermal rating of worm gears Heat dissipated	Q_g = Power lost in friction $= P(1-\eta)$ $Q_d = K'\ A\ \Delta t$ A – Area of the housing	P = Power transmitted in watts η = Efficiency of the worm gearing Conductivity of the material, K' = 378 w/m^2/^{o}C Temperature difference, Δt not to exceed 27 to 38°C

As per AGMA recommendations, limiting power of plain worm gear from heat dissipation standpoint,

$$P = \frac{3650\, x^{1.7}}{V.R + 5}$$

where

x – Centre distance in metres

P – Permissible input power in kW

V.R – Velocity ratio or transmission ratio

Forces Acting on Worm Gears

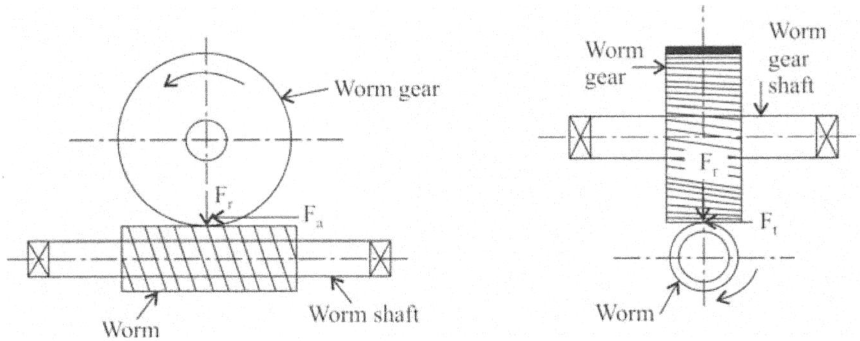

Fig. 8.3 Forces acting on worm teeth

1. Tangential force on the worm:

$$F_t = \frac{2 \times \text{Torque on worm}}{D_W} = \text{Axial force or thrust on the worm gear}$$

2. Axial force on the worm:

$$F_a = \frac{F_t}{\tan \lambda} = \frac{2 \times \text{Torque on worm gear}}{D_G}$$

3. Radial or separating force:

$$F_r = F_a \tan \phi$$

For number of starts to be used on the worm, refer Table 8.2

For proportions for worm and worm gear, refer Tables 8.3 & 8.4

Table 8.1 Values of load stress factor

Material		Load stress factor (K) N/mm^2
Worm	**Worm gear**	
Steel (B.H.N. 250)	Phosphor bronze	0.415
Hardened steel	Cast iron	0.345
Hardened steel	Phosphor bronze	0.550
Hardened steel	Chilled phosphor bronze	0.830
Hardened steel	Antimony bronze	0.830
Cast iron	Phosphor bronze	1.035

Note: The value of K given in the above table are suitable for lead angles upto 10°. For lead angles between 10° and 25°, the values of K should be increased by 25 percent and for lead angles greater than 25°, increase the value of K by 50 percent.

Table 8.2 Number of starts to be used on the worm

Velocity ratio (V.R)	36 and above	12 to 36	8 to 12	6 to 12	4 to 10
Number of starts or threads on the worm ($n = T_W$)	Single	Double	Triple	Quadruple	Sextuple

Table 8.3 Proportions for worm

Particulars	Single and double threaded worms	Triple and quadruple threaded worms
Normal pressure angle (ϕ)	$14\frac{1}{2}^\circ$	20°
Pitch circle diameter for worms integral with the shaft	$2.35\ p_c + 10$ mm	$2.35\ p_c + 10$ mm
Pitch circle diameter for worms bored to fit over the shaft	$2.4\ p_c + 28$ mm	$2.4\ p_c + 28$ mm
Maximum bore for shaft	$p_c + 13.5$ mm	$p_c + 13.5$ mm
Hub diameter	$1.66\ p_c + 25$ mm	$1.726\ p_c + 25$ mm
Face length (L_W)	$p_c (4.5 + 0.02\ T_W)$	$p_c (4.5 + 0.02\ T_W)$
Depth of tooth (h)	$0.686\ p_c$	$0.623\ p_c$
Addendum (a)	$0.318\ p_c$	$0.286\ p_c$

Note: The pitch circle diameter of the worm (D_W) in terms of the centre distance between the shafts (x) may be taken as $D_W = \frac{(x)^{0.875}}{1.416}$ (when x is in mm)

Table 8.4 Proportions for worm gear

Particulars	Single and double threads	Triple and quadruple threads
Normal pressure angle (ϕ)	$14\frac{1}{2}^\circ$	20°
Outside diameter (D_{OG})	$D_G + 1.0135\ p_c$	$D_G + 0.8903\ p_c$
Throat diameter (D_T)	$D_G + 0.636\ p_c$	$D_G + 0.572\ p_c$
Face width (b)	$2.38\ p_c + 6.5$ mm	$2.15\ p_c + 5$ mm
Radius of gear face	$0.882\ p_c + 14$ mm	$0.914\ p_c + 14$ mm
Radius of gear rim	$2.2\ p_c + 14$ mm	$2.1\ p_c + 14$ mm

Chapter 9

Mechanical Springs

9.1 Helical Springs

Table 9.1 Stress and deflection of helical springs

	Circular section	**Rectangular Section**
Shear stress	$\tau = \dfrac{8PDK}{\pi d^3}$	$\tau = \dfrac{Q_1 PD}{2bt^2}$
Deflection	$\delta = \dfrac{8PD^3 n}{Gd^4}$	$\delta = \dfrac{Q_2 \pi PD^3 n}{4Gbt^3}$
Energy stored	$E = \dfrac{1}{2} P\delta$	$E = \dfrac{1}{2} P\delta$

where

P – axial load

D – mean diameter of spring

d – diameter of wire

C – spring index, D/d

b – breadth of cross-section of wire

t – thickness of wire

n – number of active turns

p – pitch of the coil

$$= \frac{\text{free length}}{n' - 1}$$

n' – total number of turns

G – modulus of rigidity

= 0.86×10^5 N/mm^2 for carbon steels

K – wahl stress factor

$$= \frac{4C-1}{4C-4} + \frac{0.615}{C}$$

Q_1, Q_2 – factors for springs of rectangular wire section

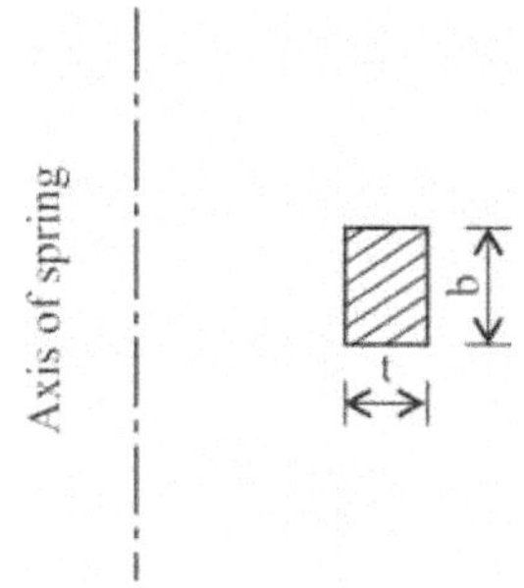

Fig. 9.1 Rectangular cross-section of wire

Table 9.2 Factors for rectangular wire section

b/t	1	1.5	2	3	4	6	8	10	∞
Q_1	4.79	4.35	4.05	3.71	3.52	3.35	3.25	3.20	3
Q_2	7.09	5.1	4.36	3.8	3.56	3.36	3.26	3.21	3

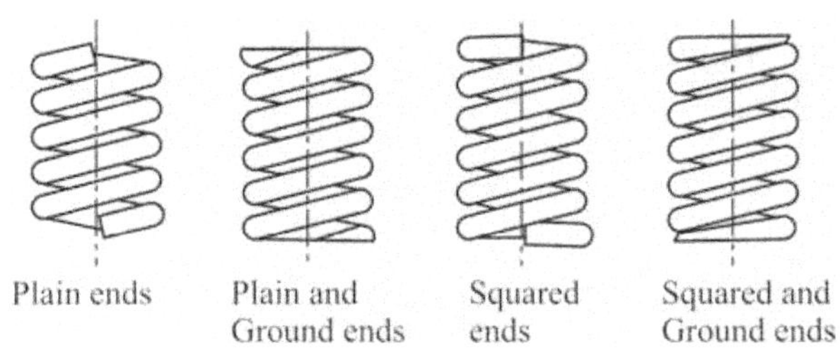

Fig. 9.2 End connections

Table 9.3 End conditions and length of springs

Type of ends	Total coils	Free length, L_f	Solid length, L_s
Plain	n	pn + d	dn + d
Plain and ground	n	pn	dn
Squared	n + 2	pn + 3d	dn + 3d
Squared and ground	n + 2	pn + 2d	dn + 2d

Natural frequency of helical compression springs

The natural frequency of helical compression springs held between two parallel plates is

$$\omega = \frac{1}{2}\sqrt{\frac{k}{m}}$$

For the case when end of the spring is on the flat plate and the other end free, supporting the external force, the natural frequency of the spring, $\omega = \frac{1}{4}\sqrt{\frac{k}{m}}$

where k – stiffness of spring (N/m)

m – mass of spring (kg)

$= Al\rho$

where A – cross-sectional area of the spring

l – length of spring

$= \pi Dn'$

ρ – density of spring material

To prevent resonance, the natural frequency of the spring should be 20 times the frequency of excitation of the external force.

Helical springs under fatigue loading

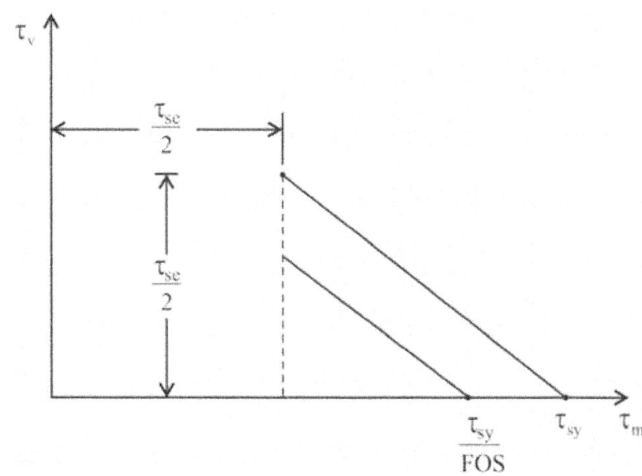

Fig. 9.3 Helical springs under fatigue loading

$$\frac{\tau_m - \tau_v}{\tau_{sy}} + \frac{2\tau_v}{\tau_{se}} = \frac{1}{FOS}$$

where

τ_m – mean shear stress

$$= \frac{8P_m D}{\pi d^3} K_s$$

K_S – shear stress multiplication factor

$$= 1 + \frac{1}{2C}$$

P_m – mean load

$$= \frac{1}{2}(P_{max} + P_{min})$$

τ_v – variable shear stress

$$= \frac{8P_v D}{\pi d^3} K$$

P_v – variable load

$$= \frac{1}{2}(P_{max} - P_{min})$$

P_{max} – maximum value of fluctuating external force on spring
P_{min} – minimum value of fluctuating external force on spring
K – wahl stress factor
τ_{sy} – yield strength in shear
τ_{se} – endurance strength in shear
FOS – factor of safety

9.2 Co-axial Springs

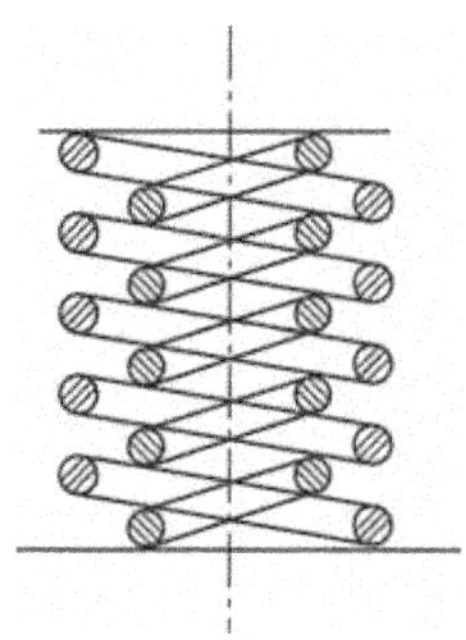

Fig. 9.4 Co-axial springs

Symbols

P – axial load

P_1 – load shared by outer spring

P_2 – load shared by inner spring

d_1 – wire diameter of outer spring

d_2 – wire diameter of inner spring

D_1 – mean coil diameter of outer spring

D_2 – mean coil diameter of inner spring

δ_1 – deflection of outer spring

δ_2 – deflection of inner spring

n_1 – number of active turns of outer spring

n_2 – number of active turns of inner spring

τ_1 – maximum shear stress in outer spring

τ_2 – maximum shear stress in inner spring

C – spring index

Assumptions

1. The springs are made of the same material and the concentric springs are designed for the same stress.
2. Both inner and outer spring coils have same spring index.
3. They have the same free length and same deflection.
4. The solid lengths of both the springs are equal.

$$\tau_1 = \tau_2$$

$$\frac{D_1}{d_1} = \frac{D_2}{d_2} = C$$

$$\delta_1 = \delta_2$$

$$n_1 d_1 = n_2 d_2$$

$$\frac{P_1}{P_2} = \frac{d_1^2}{d_2^2}$$

When the radial clearance between the two springs is taken as $\frac{(d_1 - d_2)}{2}$,

$$\frac{d_1}{d_2} = \frac{C}{C-2}$$

9.3 Torsion Springs

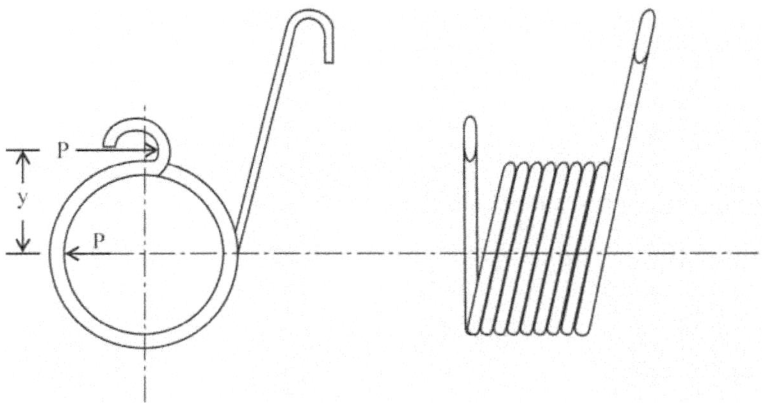

Fig. 9.5 Torsion spring

σ_b – bending stress in helical torsion spring

$$= \frac{32M}{\pi d^3} K$$

where K – stress factor

$$= \frac{4C^2 - C - 1}{4C(C-1)} \quad \text{(C – spring index)}$$

θ – angular deflection of spring (in radians)

$$= \frac{64\,MDn}{Ed^4}$$

M – bending moment acting on the spring

$= P \times y$

P – load acting on the spring

y – distance of load from the spring axis

d – diameter of spring wire

D – mean diameter of spring coil

n – number of turns

E – modulus of elasticity

For rectangular wire having width b and thickness t,

$$\sigma_b = \frac{6M}{tb^2}K$$

where $$K = \frac{3C^2 - C - 0.8}{3C^2 - 3C} \qquad \left(C = \frac{D}{b}\right)$$

$$\theta = \frac{12M\pi Dn}{Etb^3}$$

For square wire with each side equal to b,

$$\sigma_b = \frac{6M}{b^3}K$$

where $$K = \frac{3C^2 - C - 0.8}{3C^2 - 3C} \qquad \left(C = \frac{D}{b}\right)$$

$$\theta = \frac{12M\pi Dn}{Eb^4}$$

9.4 Leaf Springs

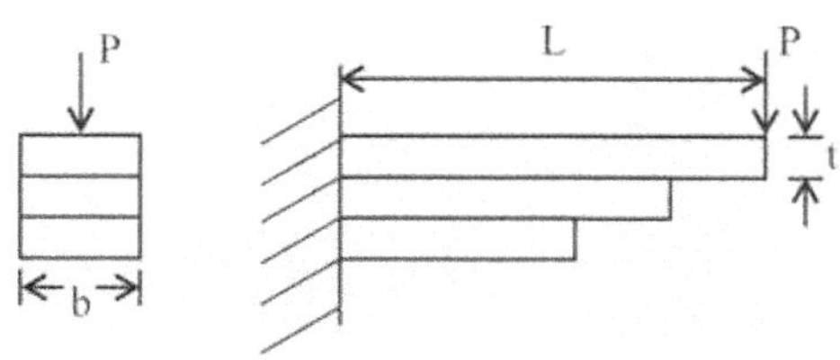

Fig. 9.6 Leaf spring

For a laminated spring (treating the spring as a cantilever beam of uniform strength),

σ_b – bending stress

$$= \frac{6PL}{nbt^2}$$

δ – deflection

$$= \frac{6PL^3}{Enbt^3}$$

where P – load on the spring

L – length of cantilever beam

n – number of leaves in the spring

b – width of the leaf

t – thickness of the leaf

E – modulus of elasticity

Semi-elliptical leaf spring

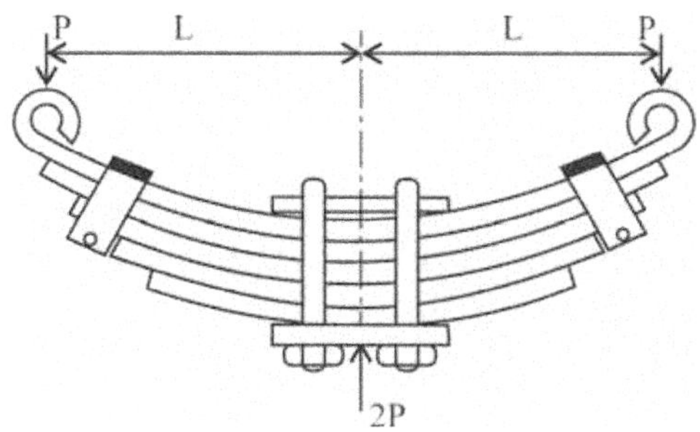

Fig. 9.7 Semi-elliptical leaf spring

σ_F – bending stress in extra full length leaves

$$= \frac{18PL}{(3n_F + 2n_G)bt^2}$$

σ_G – bending stress in graduated leaves

$$= \frac{12PL}{(3n_F + 2n_G)bt^2}$$

δ – deflection of the spring

$$= \frac{12PL^3}{(3n_F + 2n_G)Ebt^3}$$

where 2P – load on the spring

2L – length of semi-elliptical leaf spring

n_F – number of extra full length leaves

n_G – number of graduated leaves

b – width of the leaf

t – thickness of the leaf

E – modulus of elasticity

When nipping is provided in leaf springs,

σ_b – bending stress in all leaves

$$= \frac{6PL}{nbt^2}$$

where n – total number of leaves

$= n_F + n_G$

c – nip (initial gap between extra full-length leaf and the graduated length leaf)

$$= \frac{2PL^3}{Enbt^3}$$

P_i – initial pre-load required to close the gap between the extra full-length leaves and graduated length leaves

$$= \frac{2n_F n_G P}{n(3n_F + 2n_G)}$$

Chapter 10

Design of Curved Beams

Symbols

M_b – bending moment

a – area of cross section

y – distance of fibre from neutral axis

R_n – radius of curvature of neutral axis

R_o – outer radius of curvature of beam

R_i – inner radius of curvature of beam

R – distance from the centre of curvature to the centroidal axis

e – distance between the centroidal axis and the neutral axis

$= R - R_n$

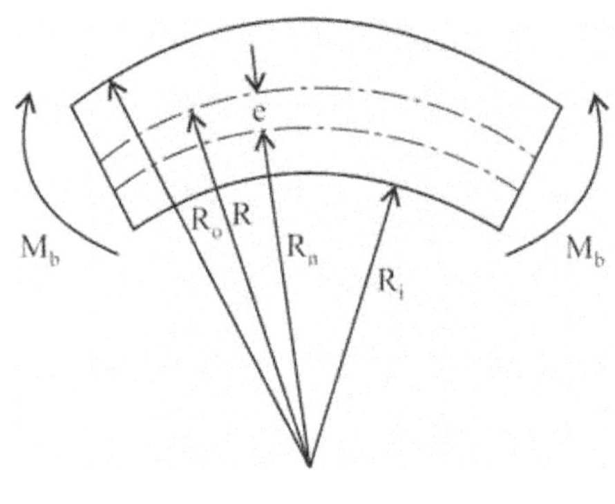

Fig. 10.1 Curved beam

$$\text{Bending stress, } \sigma_b = \frac{M_b y}{ae(R_n - y)}$$

Location of neutral axis and centroidal axis for different cross sections of curved beam:

(i) Circular section

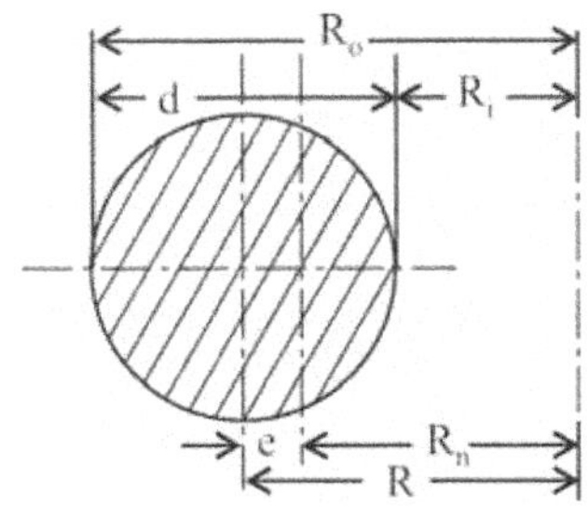

Fig. 10.2

$$R_n = \frac{\left[\sqrt{R_o} + \sqrt{R_i}\right]^2}{4}$$

$$R = R_i + \frac{d}{2}$$

(ii) Rectangular section

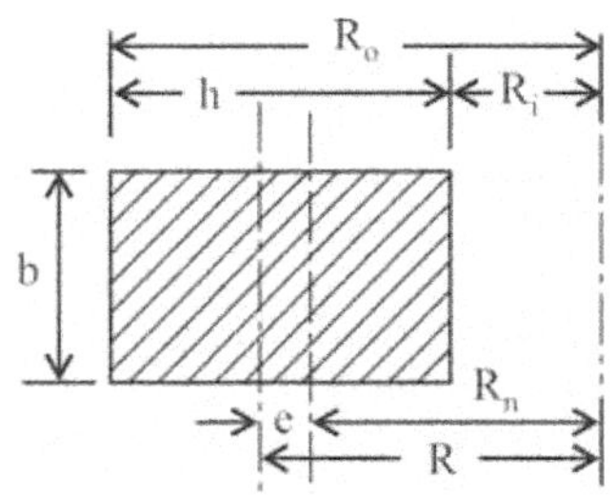

Fig. 10.3

$$R_n = \frac{h}{\ln\left(\frac{R_o}{R_i}\right)}$$

$$R = R_i + \frac{h}{2}$$

(iii) Trapezoidal section

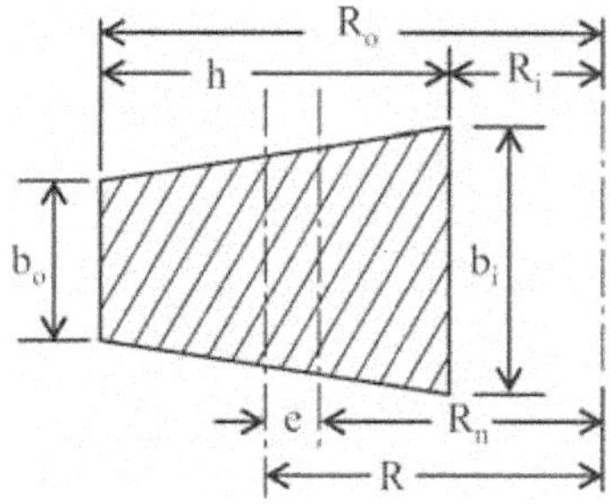

Fig. 10.4

$$R_n = \frac{\frac{1}{2}(b_i + b_o)h}{\left(\frac{b_i R_o - b_o R_i}{h}\right)\ln\left(\frac{R_o}{R_i}\right) - (b_i - b_o)}$$

$$R = R_i + \frac{h(b_i + 2b_o)}{3(b_i + b_o)}$$

(iv) I – section

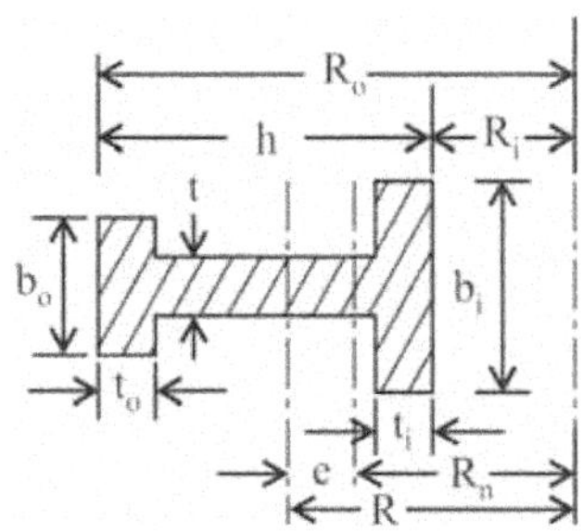

Fig 10.5

$$t_o = \frac{(b_i - t)t_i + (b_o - t)t_o + th}{b_i \ln\left(\frac{R_i + t_i}{R_i}\right) + t\ln\left(\frac{R_o - t_o}{R_i + t_i}\right) + b_o \ln\left(\frac{R_o}{R_o - t_o}\right)}$$

$$R = R_i + \frac{\frac{1}{2}h^2 t + \frac{1}{2}t_i^2 (b_i - t) + (b_o - t)t_o\left(h - \frac{1}{2}t_o\right)}{(b_i - t)t_i + (b_o - t)t_o + th}$$

(v) T–section

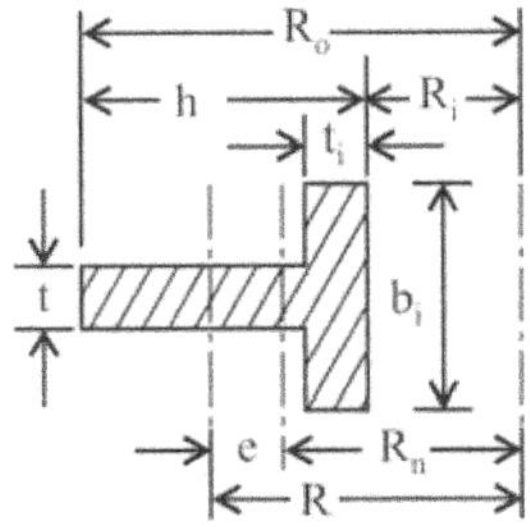

Fig. 10.6

$$R_n = \frac{(b_i - t)t_i + th}{(b_i - t)\ln\left(\frac{R_i + t_i}{R_i}\right) + t\ln\left(\frac{R_o}{R_i}\right)}$$

$$R = R_i + \frac{\frac{1}{2}h^2 t + \frac{1}{2}t_i^2 (b_i - t)}{ht + (b_i - t)t_i}$$

Chapter 11

Machine Tool Elements

11.1 Levers

Forces acting on the levers are 1. Load (W), 2. Effort (P) and 3. Reaction at the fulcrum (R).

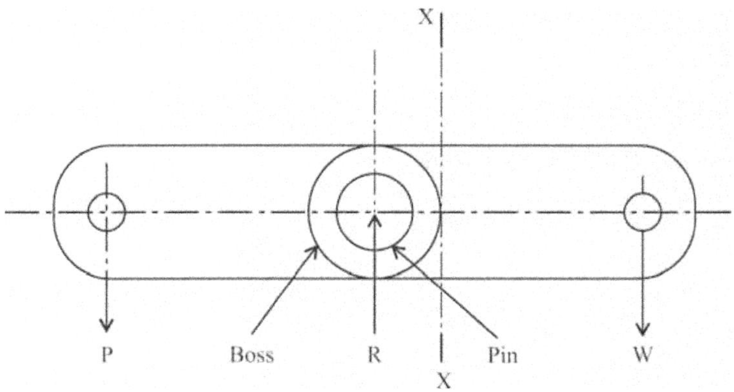

Fig. 11.1 Lever

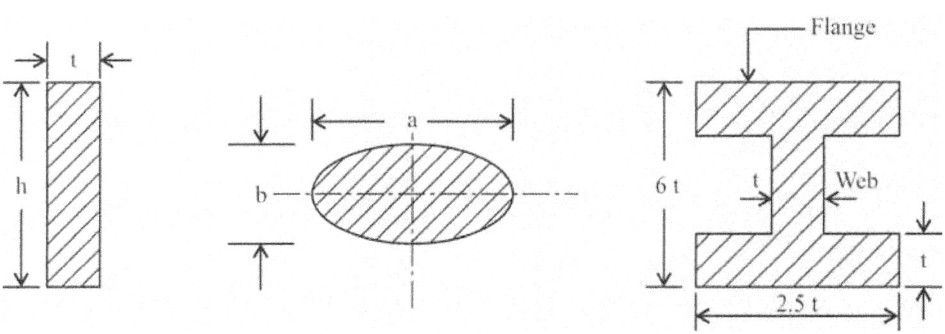

Fig. 11.2 Cross sections of lever arm (section at X-X)

Design of Cross Section of Lever

M – maximum bending moment at the boss

Section modulus for the rectangular section, $Z = \frac{1}{6} th^2$

Section modulus for the elliptical section, $Z = \frac{\pi}{32} ba^2$

Section modulus for the I-section, $Z = 12.3\ t^3$

Design of Fulcrum Pin

d – diameter of pin

l – length of pin, usually taken from 1 to 1.25 times the diameter of pin

p_b – permissible bearing pressure for the pin (5 to 10 N/mm^2)

R – reaction at the fulcrum

$= p_b \times ld$

The dimensions of the fulcrum pin are obtained from bearing consideration and then checked for shear stress.

Design of Boss

Diameter of the boss = 2d

Length of the boss = length of the pin

11.2 Hand Lever and Foot Lever

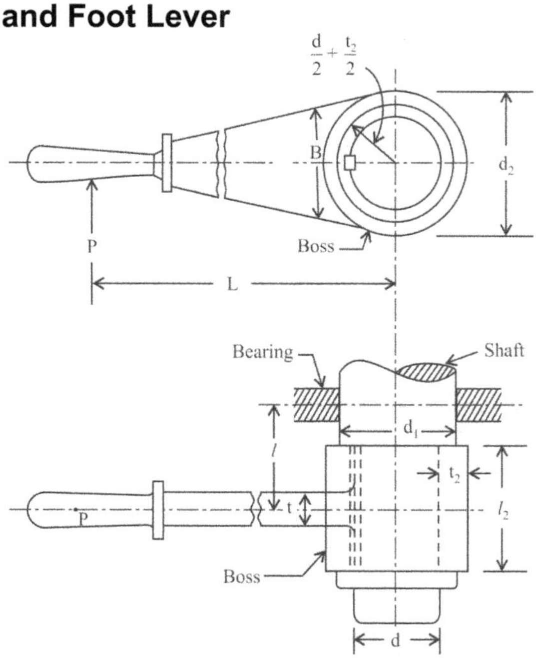

Fig. 11.3 Hand lever

P – force applied

L – effective length

σ_t, τ – permissible tensile and shear stresses respectively

(i) Diameter 'd' obtained on pure torsion;

$$T = P \times L$$

$$T = \frac{\pi}{16}\tau d^3$$

$$\therefore \quad d = \sqrt[3]{\frac{16T}{\pi\tau}}$$

(ii) Diameter of boss 'd_2'

$$d_2 = 1.6\ d$$

Thickness of boss, $t_2 = 0.3d$

(iii) Length of the boss, $l_2 = d$ to $1.25\ d$

$$l_2 = \frac{2P \times L}{t_2\sigma_t(d + t_2)}$$

(iv) Diameter of the shaft at the center of bearing (d_1)

$$T_e = \sqrt{M^2 + T^2} = \sqrt{(Pl)^2 + (PL)^2} = P\sqrt{l^2 + L^2} = \frac{\pi}{16}\tau(d_1)^3$$

The length l may be taken as $2l_2$.

(v) Key for the shaft is designed for transmitting a torque of P × L.

$$\frac{w}{t} = \frac{\sigma_c}{2\tau}$$

where w – width of the key

t – thickness of the key

σ_c – permissible crushing stress for the key material

τ – permissible shear stress for the key material

(vi) The cross section of the lever near the boss is determined by considering the lever in bending.

$$\sigma_b = \frac{6PL}{t \times B^2}$$

where σ_b – bending stress in the lever

t – thickness of lever near the boss

B – width of lever near the boss
= 4t to 5t

The width of the lever near the handle is $\frac{B}{2}$.

A foot lever is similar to hand lever and it may be designed in a similar way as discussed for hand lever.

11.3 Cranked Lever

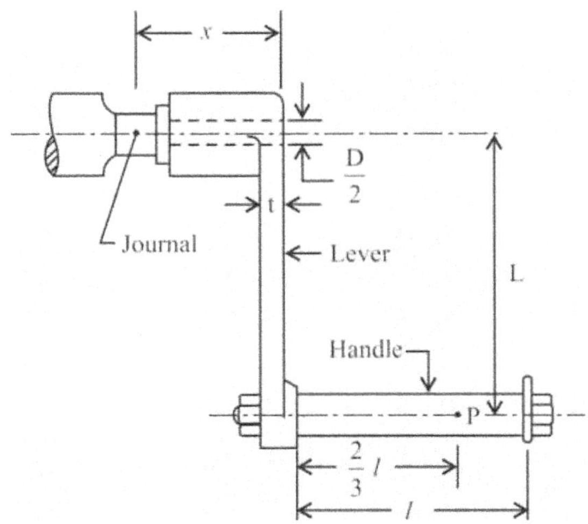

Fig. 11.4 Cranked lever

(i) Diameter of handle

Maximum bending moment, $M = \frac{2}{3} P \times l$

Section modulus, $Z = \frac{\pi d^3}{32}$

where d – diameter of handle

$$\sigma_b \frac{\pi}{32} d^3 = \frac{2}{3} P \times l$$

$$d = \sqrt[3]{\frac{64 P l}{3 \pi \sigma_b}}$$

(ii) Cross section of lever arm

Generally maximum bending moment, M = 1.25 P L

Section modulus for the lever arm, $Z = \frac{1}{6} tB^2$

$$\sigma_b = \frac{M}{Z}$$

t – thickness

B – width of the lever arm near the boss

B is taken as twice the thickness.

(iii) Check for induced shear stress in the section of the lever arm near the boss

$$T = \frac{2}{9} Bt^2 \times \tau$$

(iv) Check for maximum principal and shear stresses

$$\sigma_{b(max)} = \frac{1}{2}\left[\sigma_b + \sqrt{(\sigma_b)^2 + 4\tau^2}\right]$$

$$\tau_{max} = \frac{1}{2}\sqrt{(\sigma_b)^2 + 4\tau^2}$$

(v) Diameter of journal

$$D = \sqrt[3]{\frac{16T_e}{\pi\tau}}$$

where $T_e = P\sqrt{\left(\frac{2l}{3} + x\right)^2 + L^2}$

x – distance from the end of boss to the centre of journal

11.4 Lever for a Lever Safety Valve

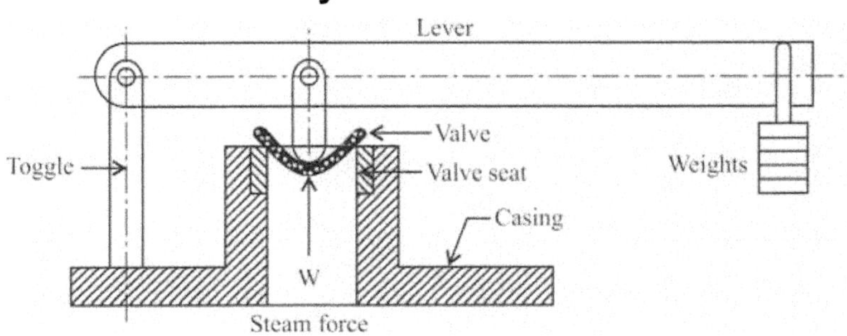

Fig.11.5 Lever safety valve

Maximum steam load, $W = \frac{\pi}{4} D^2 p$

p – steam pressure

D – diameter of the valve

Lever to be designed as discussed earlier.

11.5 Bell Crank Lever

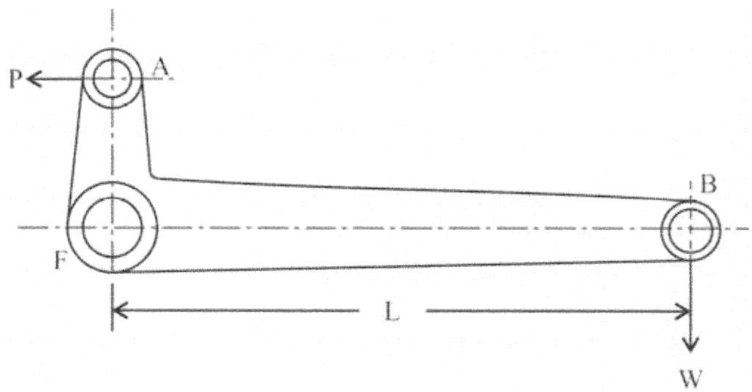

Fig. 11.6 Bell crank lever

$$R = \sqrt{W^2 + P^2}$$

(i) Design of fulcrum pin

$$R = d \times l \times p_b$$

$$l = 1.25\ d$$

Allowable bearing pressure p_b may be taken as 10 N/mm^2.

Check for induced shear stress in the fulcrum pin.

(ii) Design of pin at A

Generally same values are adopted as for the fulcrum pin.

(iii) Design of pin at B

$$W = d_1 \times l_1 \times p_b$$

$$l_1 = 1.25\ d_1$$

(iv) Design of lever

$$M = W \times L$$

$$Z = \frac{1}{6} t b^2$$

$$b = 3t$$

t – thickness of the lever

b – width of the lever

Check for the induced bending stress in the lever.

11.6 Rocker Arm for Exhaust Valve

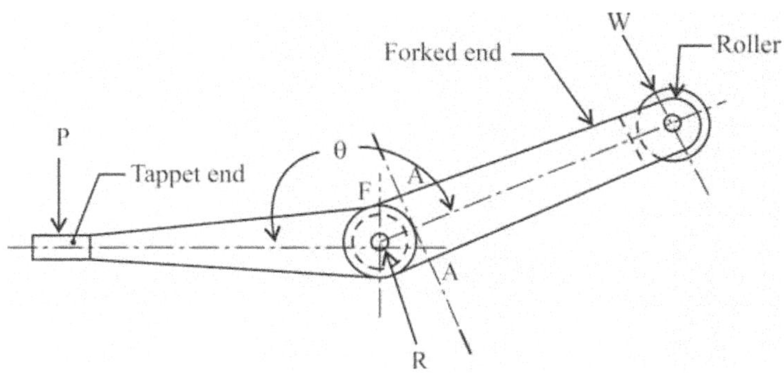

Fig. 11.7 Rocker arm for exhaust valve

$$R = \sqrt{W^2 + P^2 - 2WP\cos\theta}$$

(i) Design of fulcrum

$l = 1.25d$

$R = d \times l \times p_b$

Check for shear stress in the pin, $\tau = \dfrac{2R}{\pi d^2}$

External diameter of boss, $D = 2d$

(ii) Design for forked end

Diameter of the roller pin = d_1

Length of the roller pin, $l_1 = 1.25d_1$

Load on the roller pin, $W = l_1 \times d_1 \times p_b$

Check for induced shear stress.

$$W = 2 \times \frac{\pi}{4} d_1^2 \times \tau$$

Check for bending stress in the roller pin.

Maximum bending moment, $M = \frac{5}{24} W \times l_1$

Section modulus of the pin, $Z = \frac{\pi}{32} d_1^3$

Bending stress induced in the pin $= \frac{M}{Z}$

(iii) Design of lever arm

$$\sigma_b = \frac{M}{Z}$$

M = bending moment at section A – A

Z = section modulus

(iv) Design for tappet screw

d_c – core diameter of screw

Load on the tappet screw, $P = \frac{\pi}{4} d_c^2 \sigma_c$

σ_c – allowable compressive stress

= 50 N/mm^2 for mild steel.

Nominal diameter of the screw, $d = \frac{d_c}{0.84}$

Chapter 12

Shafts and Axles

Symbols

P – Power transmitted, watts

N – Speed, rpm

d – Diameter of shaft, mm

d_o – Outer diameter of hollow shaft, mm

k – Ratio of inner diameter to outer diameter of hollow shaft

σ_b – Allowable bending stress, N/mm^2

τ – Allowable shear stress, N/mm^2

M – Bending moment, N-mm

M_e – Equivalent bending moment, N-mm

T – Twisting moment, N-mm

T_e – Equivalent twisting moment, N-mm

F – Axial load, N

K_m – Combined shock and fatigue factor for bending

K_t – Combined shock and fatigue factor for torsion

L – Length of shaft between the bearings, mm

K – Least radius of gyration, mm

θ – Torsional deflection, rad

g – Acceleration due to gravity, m/sec^2

ω_c – Critical speed, rad/sec

I – Moment of inertia, mm^4

J – Polar moment of inertia of the cross-sectional area, mm^4

E – Modulus of elasticity, N/mm^2

G – Modulus of rigidity for the shaft material, N/mm^2

12.1 Design Equations

Twisting moment, $T = \dfrac{P \times 60}{2\pi N}$

Shafts subjected to twisting moment

$$d = \sqrt[3]{\frac{16T}{\pi\tau}} \quad \text{for a solid shaft}$$

$$d_o = \sqrt[3]{\frac{16T}{\pi\tau(1-k^4)}} \quad \text{for a hollow shaft}$$

For belt drives, twisting moment $T = (T_1 - T_2)\,R$

where T_1 T_2 – Tensions in the tight and slack sides of the belt respectively, N

R – Radius of the pulley, mm

Shafts subjected to bending moment

$$d = \sqrt[3]{\frac{32M}{\pi\sigma_b}} \quad \text{for a solid shaft}$$

$$d = \sqrt[3]{\frac{32M}{\pi\sigma_b(1-k^4)}} \quad \text{for a hollow shaft}$$

For an axle, $d = \sqrt[3]{\dfrac{32M}{\pi\sigma_b}}$

Shafts subjected to combined twisting moment and bending moment

$$d = \sqrt[3]{\frac{16T_e}{\pi\tau}} \quad \text{for a solid shaft}$$

$$d_o = \sqrt[3]{\frac{16T_e}{\pi\tau(1-k^4)}} \qquad \text{for a hollow shaft}$$

where $$T_e = \sqrt{M^2 + T^2}$$

$$d = \sqrt[3]{\frac{32M_e}{\pi\sigma_b}} \qquad \text{for a solid shaft}$$

$$d_o = \sqrt[3]{\frac{32M_e}{\pi\sigma_b(1-K^4)}} \qquad \text{for a hollow shaft}$$

where $$M_e = \frac{1}{2}\left[M + \sqrt{M^2 + T^2}\right]$$

When combined shock and fatigue factors are considered, the expressions for T_e and M_e may be replaced as

$$T_e = \sqrt{(K_m M)^2 + (K_t T)^2} \qquad \text{and}$$

$$M_e = \frac{1}{2}\left[K_m M + \sqrt{(K_m M)^2 + (K_t T)^2}\right]$$

For values of K_m and K_t, refer Table 12.1.

Shafts subjected to axial compression load in addition to combined twisting moment and bending moments

$$d = \sqrt[3]{\frac{16T_e}{\pi\tau}} \qquad \text{for a solid shaft}$$

where $$T_e = \sqrt{\left[K_m M + \frac{\alpha F d}{8}\right]^2 + (K_t T)^2}$$

$$d_o = \sqrt[3]{\frac{16T_e}{\pi\tau(1-k^4)}} \qquad \text{for a hollow shaft}$$

where $$T_e = \sqrt{\left[K_m M + \frac{\alpha F d_o(1+k^2)}{8}\right] + (K_t T)^2}$$

$$d = \sqrt[3]{\frac{32M_e}{\pi\sigma_b}} \qquad \text{for a solid shaft}$$

where $$M_e = \frac{1}{2}\left[K_m M + \frac{\alpha F d}{8}\right] + \sqrt{\left\{K_m M + \frac{\alpha F d}{8}\right\}^2 + (K_t T)^2}$$

$$d = \sqrt[3]{\frac{32 M_e}{\pi \sigma_b (1 - k^4)}}$$ for a hollow shaft

where $$M_e = \frac{1}{2}\left[K_m M + \frac{\alpha F d_o (1 + k^2)}{8} + \sqrt{\left\{K_m M + \frac{\alpha F d_o (1 + k^2)}{8}\right\}^2 + (K_t T)^2}\right]$$

Column action factor, $\alpha = \frac{1}{1 - 0.0044(L/K)}$ when the slenderness ratio (L/K) is less than 115.

When (L/K) is more than 115, $\alpha = \frac{\sigma_y (L/K)^2}{C \pi^2 E}$

where σ_y – Compressive yield point stress of shaft material and

C – Coefficient in Euler's formula depending upon the end conditions

= 1 for hinged ends

= 2.25 for fixed ends

= 1.6 for ends that are partly restrained as in bearings

For axial tensile load, $\alpha = 1$

Design of shafts on the basis of torsional rigidity

$$\theta = \frac{T \times L}{G \times J}$$

Table 12.1 Recommended values for K_m and K_t

Type	K_m	K_t
Stationary shafts		
Gradually applied load	1	1
Suddenly applied load	1.5-2	1.5-2
Revolving shafts		
Gradual loading	1.5	1
Minor shock loads	1.5-2	1-1.5
Heavy shock loads	2-3	1.5-3

12.2 Deflections and Critical Speeds of Shafts under various loading conditions

A shaft of negligible self-weight with single attached disc

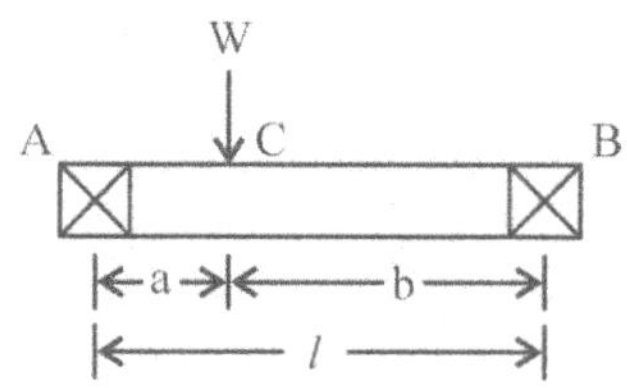

Fig. 12.1 A shaft with single load

δ_c – Deflection at point C

$$= \frac{Wa^2b^2}{3EIl}$$

$$\omega_c = \sqrt{\frac{g}{\delta_c}} \qquad \text{where } g = 9.81 \text{ m/s}^2$$

A shaft with self-weight only

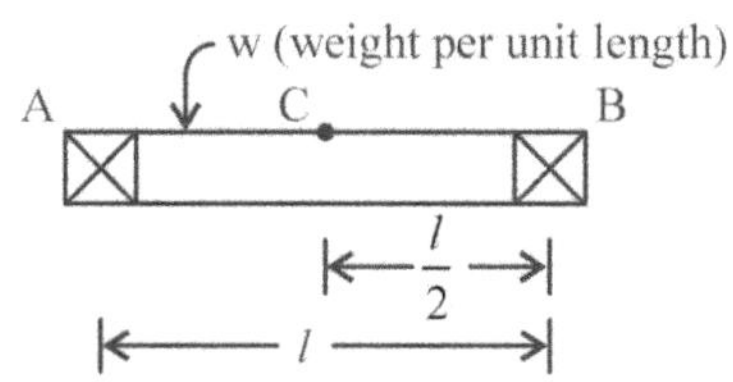

Fig. 12.2 A shaft with self-weight only

δ_{c_s} – Deflection at point C due to self-weight only

$$= \frac{5wl^4}{384EI}$$

$$\omega_c = \sqrt{\frac{5}{4}\left(\frac{g}{\delta_{c_s}}\right)} \qquad \text{where } g = 9.81 \text{ m/s}^2$$

A shaft having self-weight and carrying several concentrated disc weights

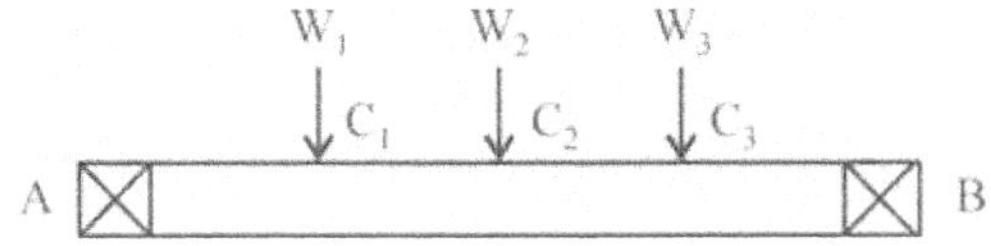

Fig. 12.3 A shaft with self-weight and several concentrated loads

The deflections δ_{c_1}, δ_{c_2}, δ_{c_3} are obtained for each load acting individually and δ_{c_s} is obtained for self-weight individually to determine the corresponding critical speeds ω_{c_1}, ω_{c_2}, ω_{c_3} and ω_{c_s}.

ω_c, the critical speed of shaft is determined from the equation,

$$\frac{1}{\omega_c^2} = \frac{1}{\omega_{c_1}^2} + \frac{1}{\omega_{c_2}^2} + \frac{1}{\omega_{c_3}^2} + \frac{1}{\omega_{c_s}^2} \qquad \text{(Dunkerley equation)}$$

Critical speed in rpm, $N_c = \dfrac{30\ \omega_c}{\pi}$

The shaft speed has to be 20% away from the critical or whirling speed of shaft to avoid resonance.

Appendix

Table 1 Properties of materials

Material	Mass density kg/m^3	Modulus of elasticity (E) N/mm^2	Modulus of rigidity (G) N/mm^2	Poisson's ratio
Aluminium	2700	0.675×10^5	0.26×10^5	0.34
Beryllium	1820	2.928×10^5	0.26×10^5	0.34
Brass	8450	0.97×10^5	0.35×10^5	0.30 – 0.40
Bronze	8730	1.11×10^5	0.35×10^5	0.30 – 0.40
Cast iron	7200	1.00×10^5	0.35×10^5	0.23
Copper	8960	1.23×10^5	0.39×10^5	0.26
Lead	11340	0.16×10^5	0.076×10^5	0.45
Monel metal	8580	1.59×10^5	0.67×10^5	0.32
Carbon steel	7800	2.07×10^5	0.8×10^5	0.30
Titanium	4540	1.05×10^5	0.43×10^5	0.33
Tungsten	19300	4.153×10^5	1.77×10^5	0.17

Table 2 Properties of carbon steels

Designation	Tensile strength N/mm^2	Yield strength N/mm^2	Brinell hardness number (HB)
C 07	314-392	196	-
C 10	333-412	206	-
C 14	363-441	216	137
C 15	363-481	235	137
C 15 Mn 75	412-490	245	163
C 20	432-510	245	156
C 25	432-520	275	170
C 25 Mn 75	461-560	275	207
C 30	490-588	294	179
C 35	510-608	304	187
C 35 Mn 75	540-637	314	223
C 40	570-667	324	217
C 45	618-696	353	229
C 50	647-765	373	241
C 50 Mn 1	706 min	392	255
C 55 Mn 75	706 min	392	265
C 60	736 min	412	255
C 65	736 min	422	255

Table 3 Properties of alloy steels

Designation	Tensile strength N/mm^2	Yield strength N/mm^2	Brinell hardness number (HB)
20 Mn 2 and	588-736	432	170-217
27 Mn 2	687-834	490	201-248
37 Mn 2	588-736	432	170-217
	687-834	530	201-248
	785-932	588	229-277
	882-1030	687	255-311
35 Mn 2 Mo 28	687-834	530	201-248
	785-932	588	229-277
	882-1030	687	255-311
	981-1128	785	285-341
35 Mn 2 Mo 45	785-932	588	229-277
	882-1030	687	255-311
	981-1128	785	285-341
40 Crl Mo 28	687-834	530	201-248
	785-932	588	229-277
	882-1030	687	255-311
	981-1128	785	285-341
15 Cr3 Mo 55 and 25 Cr3 Mo 55	687-834	530	201-248
	785-932	588	229-277
	882-1030	687	255-311
	981-1128	785	285-341
	1080-1226	863	311-363
	1520 min.	1275	444 min.
40 Cr 3 Mol V 20	1324 min.	1098	363 min.
	1520 min.	1275	448 min.
40 Cr 2 All Mo 18	687-834	530	201-248
	785-932	588	229-277
	822-1030	687	255-311
40 Ni 3	785-932	588	229-277
	822-1030	687	255-311
35 Ni l Cr 60	687-834	530	201-248
	785-932	588	229-277
	882-1030	687	255-311
30 Ni 4 Cr 1	1520 min.	1275	444 min.
40 Ni 2 Cr 1 Mo 15	785-932	588	229-277
	882-1030	687	255-311
	981-1128	785	285-431
	1080-1226	863	311-363
40 Ni 2 Cr 1 Mo 28	785-932	588	229-277
	882-1030	687	255-311
	1080-1128	785	255-341
	1080-1226	863	311-363
	1180-1324	981	341-401
	1520 min.	1275	444 min.
40 Ni 2 Cr 65 Mo 55	981-1128	785	285-341
	1080-1226	863	311-363
	1180-1324	981	341-401
	1520 min.	1275	444 min.

References

1. Design Data Book by Faculty of Mechanical Engineering, PSG College of Technology, Coimbatore, Published by M/s DPV PRINTERS, Coimbatore, 2000.
2. Design Data Handbook by S. Md. Jalaludeen, Published by Anuradha Publishers, Kumbakonam, 2004.
3. Machine Design Data Book by V.K. Jadon & Suresh Verma, Published by I.K. International Pvt. Ltd., New Delhi, 2007.
4. A Textbook of Machine Design by R.S. Khurmi & J.K. Gupta, Published by Eurasia Publishing House (Pvt.) Ltd, New Delhi, 1979.
5. Machine Design by P.C. Sharma & D.K. Agarwal, Published by S.K. Kataria & Sons, Delhi, 2009.
6. Machine Design Fundamentals and Applications by P.C. Gope, Published by PHI Learning Pvt. Ltd., New Delhi, 2012.
7. Design of Machine Elements by M.F. Spotts, T.E. Shoup and L.E. Hornberger, Published by Pearson Prentice Hall, 2004.
8. IS: 4694-1968 Basic Dimensions for Square Threads, Published by Bureau of Indian Standards.
9. Machine Design by S.G.Kulkarni, Tata McGraw Hill, 2008.
10. Design Data Handbook by K. Mahadevan and K. Balaveera Reddy, CBS Publishers and Distributors, 1987.
11. Machine Design Data Book by V. B. Bhandari, McGraw Hill Education (India) Private Limited, 2014.
12. Machine Design by P. Kannaiah, Scitech Publications (India) Private limited, 2003.

13. Machine Design by T. V. Sundara Raja Moorthy and N. Shanmugam, Anuradha Agencies, 2000.
14. Machine Design by N. C. Pandya and C. S. Shah, Charotar Publishing House Pvt. Ltd., 2014.
15. Data Book for Designing Machine Elements by Arun Kumar, S. K. Kataria and Sons, 2008.

Index

www.ingramcontent.com/pod-product-compliance
Ingram Content Group UK Ltd.
Pitfield, Milton Keynes, MK11 3LW, UK
UKHW060019200726
13845UKWH00046B/90

9 789352 301119